D0880913

Metabolic Arrest and the Control of Biological Time

Metabolic Arrest and the Control of Biological Time

Peter W. Hochachka and Michael Guppy

Harvard University Press
Cambridge, Massachusetts, and London, England 1987

Printed in the United States of America
10 9 8 7 6 5 4 3 2 1

This book is printed on acid-free paper, and its binding materials have been chosen for strength and durability.

Library of Congress Cataloging-in-Publication Data

Hochachka, Peter W.
Metabolic arrest and the control of biological time.

Bibliography: p.
Includes index.
1. Dormancy (Biology) 2. Metabolism.
I. Guppy, Michael. II. Title.
QH523.H63 1987 591.1'33 86-14860
ISBN 0-674-56976-8 (alk. paper)

Designed by Gwen Frankfeldt

To our teachers

E. T. Nepstad		P. J. Fullagar
R. P. Bishop	AND	C. Bryant
E. C. Mathews		M. J. Weidemann

whose influence was never constrained by time

Preface

Some time ago, one of us (PWH, while on leave at Monash University) brought home to his children a tiny marsupial, thinking it might be educational for them. The pygmy possum was easy and fun to handle and most interesting in its activities; thus it quickly became a favorite family pet. But it was an unusual little beast, for each morning the children would find it curled up in a tight little ball, torpid. The children readily worked out that entering torpor was the pygmy possum's way of defending himself against the cold nights that are characteristic of autumn in Melbourne. When, in two subsequent "accidents," the children discovered that their little friend would also switch off when he ran out of food or out of water, we all (including MG) became interested in the process, for we were acutely aware of the use of similar metabolic arrest strategies by other animals to extend their hypoxia tolerance. We began to wonder how extensive metabolic arrest was as a defense against harsh environmental conditions or as a strategic retreat from them. What mechanisms were utilized to turn down or to turn off cell metabolism and other cell functions? How did animals like the pygmy possum "know" when to activate mechanisms for arresting tissue and organ functions? How were these mechanisms activated? How did organisms "know" when to switch back on and terminate the arrested state? Were universal principles of metabolic arrest decipherable from available data? For the children, these kinds of questions arose from an inherent and unquenchable curiosity; for the adults, they were questions that arose out of a fundamental interest in metabolic regulation, and it is from the perspective of the latter discipline, as metabolic biochemists, that we decided to analyze the uses and mechanisms of metabolic arrest in the animal kingdom. The lessons learned, we believe, are potentially of great interest to biochemistry, biology, and medicine.

Although much of the literature relevant to our analysis of metabolically arrested states of animals is covered in detail in the text, three earlier overviews should be brought to the attention of the reader. These are *Depressed Metabolism*, ed. X. J. Musacchia and J. F. Saunders (New York: American Elsevier, 1969); Perspectives on the biology of dormancy, *Trans. Amer. Microscop. Soc.* 93:459–631 (1974), and *Dormancy and Developmental Arrest*, ed. M. E. Clutter (New York: Academic Press, 1978).

Projects that aim at synthesis and the generation of universal principles tax mental and physical energies. All the more significant, therefore, is the support and the encouragement received during our work. PWH wishes to extend special thanks to the Department of Science and Technology, Australia, for support through a Queen Elizabeth II Senior Fellowship and to Jodi Simpson of Harvard University Press for good judgment and good spirits. Moreover, this book would not have been possible without the ever-present enthusiasm (for adventure) and joy (on its experience) emanating from the family, Brenda, Claire, Gail, and Gareth Vasilii (not to mention Panda and Stanley). MG wishes to thank the staff of three libraries: the Life Sciences Library, at Australian National University, and the Biological Sciences and the Zoology libraries at the University of Western Australia. The cheerful, tireless, and expert help from these people made the writing of his chapters both possible and pleasant. PWH and MG both thank E. C. Slater for permission to reproduce the photograph of a pygmy possum appearing at the beginning of this preface.

Contents

Abbreviations

Common Metabolites

AMP; ADP; ATP	Adenosine 5′-mono-; -di-; -triphosphate
cAMP	3′,5′-Cyclic AMP
ArgP	Arginine phosphate
CMP; CDP; CTP	Cytidine 5′-mono-; -di-; -triphosphate
CoA	Coenzyme A
Cr; CrP	Creatine; creatine phosphate
DG	Diglyceride
DHAP	Dihydroxyacetone phosphate
DNA	Deoxyribonucleic acid
1,3DPG (or DPG)	1,3-Diphosphoglycerate
FAD^+; FADH	Flavin adenine dinucleotide; its reduced form
FFA	Free fatty acids
F6P	Fructose 6-phosphate
F1,6BP	Fructose 1,6-bisphosphate
F2,6BP	Fructose 2,6-bisphosphate
G1P	Glucose 1-phosphate
G3P (or GAP)	Glyceraldehyde 3-phosphate
GMP; GDP; GTP	Guanosine 5′-mono-; -di-; -triphosphate
imid	Imidazole
IMP; IDP; ITP	Inosine 5′-mono-; -di-; -triphosphate
IP_3	Inositol triphosphate
2KGA	Ketoglutarate
MG	Monoglyceride
NAD^+; NADH	Nicotinamide adenine dinucleotide; its reduced form
$NADP^+$; NADPH	Nicotinamide adenine dinucleotide phosphate; its reduced form
OXA	Oxaloacetate
PAH	Paraaminohippuric acid

PC	Phosphatidylcholine
P5C	Pyrroline 5-carboxylate
PE	Phosphatidylethanolamine
PEP	Phosphoenolpyruvate
PGA	Phosphoglycerate
PI	Phosphatidylinositol
PI(4)P	Phosphatidylinositol 4-phosphate
$PI(4,5)P_2$	Phosphatidylinositol 4,5-bisphosphate
P_i	Inorganic phosphate
PP_i	Inorganic pyrophosphate
RNA; mRNA; tRNA	Ribonucleic acid; messenger RNA; transfer RNA
TG	Triglyceride
UMP; UDP; UTP	Uridine 5′-mono; -di-; -triphosphate

Common Enzymes

CPK	Creatine phosphokinase
CS	Citrate synthase
DNase	Deoxyribonuclease
FBPase	Fructose 1,6-bisphosphatase
GDH	Glutamate dehydrogenase
α-GPDH	α-Glycerophosphate dehydrogenase
GOT	Glutamate–oxaloacetate transaminase
G6PDH	Glucose-6-phosphate dehydrogenase
GPT	Glutamate–pyruvate transaminase
HK	Hexokinase
IDH	Isocitrate dehydrogenase
2KGDH	2-Ketoglutarate dehydrogenase
LDH	Lactate dehydrogenase
MDH	Malate dehydrogenase
ME	Malic enzyme
Na^+,K^+-ATPase	Na^+,K^+-activated adenosine triphosphatase
ODH	Octopine dehydrogenase
PDH	Pyruvate dehydrogenase
PEPCK	Phosphoenolypyruvate carboxykinase
PFK	Phosphofructokinase
6PGDH	6-Phosphogluconate dehydrogenase
PGK	Phosphoglycerate kinase
PK	Pyruvate kinase

Common Terms

α_{Im}	Dissociation ratio of imidazole groups
ACR	Air convection requirement
BPM	Beats per minute

DA	Descending aorta
DPM	Disintegrations per minute
ECF	Extracellular fluid
EEG	Electroencephalogram
ETS	Electron transport system
H_{II} phase	Hexagonal II phase
Hb	Hemoglobin
ICF	Intracellular fluid
K_m	Michaelis-Menten constant
M	Body weight (mass)
mTAL	Medullary thick ascending limb
M_s	Mass of the rapidly mixing pool of any given metabolite
MW	Molecular weight
NMR	Nuclear magnetic resonance
P_{50}	O_2 tension for 50% saturation of hemoglobin
pN	pH at which $[H^+] = [OH^-]$
RBC	Red blood cells
RQ	Respiratory quotient
SA	Specific radioactivity (expressed in DPM per micromole)
SR	Sarcoplasmic reticulum
T_b	Body temperature
TEF	Time extension factor
VL	Vital limit of water loss

Rate Functions

BMR	Basal metabolic rate
$CMRO_2$	Brain metabolic rate (μmol O_2 $g^{-1}min^{-1}$)
GFR	Glomerular filtration rate
Q_{10}	Change in the rate of a process over a 10°C change in temperature
R_a, R_d	Entry rate, exit rate of metabolites into and out of the plasma
SMR	Standard metabolic rate
$T_{1/2}$	Time to 50% clearance
$\dot{V}_{O_2}$	O_2 consumption rate (usually expressed as milliliters of O_2, or as micromoles O_2 per gram per minute)

Metabolic Arrest and the Control of Biological Time

1

The Time Extension Factor

Laymen and scientists alike have frequently contemplated the ramifications and usefulness of reversible metabolic arrest, of being able to reversibly switch down or even switch off metabolism for variable and controllable time periods. The seductiveness of the idea of reversible metabolic arrest stems from the desire to escape from time. Time in biological systems can be expressed by scientists in units of standard clocking devices, but for cells, tissues, organs, and organisms time is measured with molecular or biological rate processes: the slower these rates, the greater the apparent extension of metabolic time with respect to clock time. For example, identical metabolic or biological processes take 20 times longer in elephants, and 7 times longer in men, than in mice (Table 1.1); because metabolism slows down with size, 1 sec to a mouse means 7 sec to man and 20 sec to an elephant (see Schmidt-Nielsen, 1984; Lindstedt and Calder, 1981). Slowing down molecular or metabolic rate processes therefore means slowing down or extending biological time (with respect to clock time), and a complete but reversible cessation of metabolism means a reversible escape from time. By inducing partial metabolic arrest, surgeons gain many advantages, but probably their biggest gain is time—operating time. Science fiction writers and movie makers assume the same strategy for every imaginary space odyssey for the same reason—to circumvent the constraints of time. But most surgeons and biomedical scientists realize that in complex organs and organisms metabolism can never be completely switched off and that partially arrested metabolic states *even in principle* cannot be extended for too long before they become irreversible. In fact, it is assumed (for theoretical reasons) that metabolic arrest can never be complete and can only be sustained for short time periods. This assumption, however, is contradicted by numerous organisms that routinely rely upon metabolic arrest as a reliable survival strategy. By "putting time on hold" (slowing down or stopping metabolism), some organisms are able to survive (in some cases, indefinitely) environmental conditions that would otherwise be too stressful to manage. Such animals have developed mechanisms to deal with limiting levels of three major environmental components: water, heat, and oxygen. Thus variable degrees of metabolic arrest down to and including fully ametabolic states are common strategies for dealing with problems of dessication, cold, and anoxia (that is, for dealing with the *lack* of adequate water, heat, or O_2 in the environment). At this time, fundamental mechanisms underlying metabolic arrest capabilities are gradually being worked out, and these studies form the basis for most of this book. It is of course too early to know which, if any, of these metabolic mechanisms will be transferable to applied (clinical) settings. However, we will begin our analysis by inquiring why biomedical scientists generally assume that metabolic arrest can at best be partial and why, even at these levels, it cannot be sustained for long. The answer is to be found in the widespread acceptance of the concept of maintenance metabolism. That concept states that for every organism there is some minimal rate of metabolism requisite for life and below

Table 1.1 Biological time extension and body mass

Species	Mass (kg)	Equivalent time[a]
Whale	100,000	85
Elephant	4,000	35
Horse	700	22.5
Man	70	12.5
Rat	0.2	3.0
Mouse	0.03	1.8
Shrew	0.003	1

Source: Based on Lindstedt and Calder (1981) and Schmidt-Nielsen (1984).

a. Standardized to one unit of clock time in the shrew. Because biological time varies 1/metabolic rate, homologous molecular processes or cycles occur at slower and slower rates as body mass increases.

which the system cannot operate for any extended time period. Obviously this concept appears to be contradicted by organisms capable of profound and extended metabolic arrest; what, then, is the problem of maintenance metabolism all about and how can some organisms seemingly turn it up or down with ready facility? What is maintenance metabolism?

Basal Metabolic Rate

For any given organism there is assumed to be a minimum metabolic rate required for simple maintenance of tissues. The basal metabolic rate (BMR) in human beings is considered an approximation of such maintenance metabolism and is defined as the rate at which O_2 is consumed (or the rate at which an equivalent amount of heat is produced) by an awake individual lying at rest following at least 12 hr of fasting. For a 70-kg man, BMR is about 12 mmol O_2 min^{-1} or about 0.17 mmol O_2 kg^{-1} min^{-1}. Many factors (for example, age, sex, climatic conditions, physical training, drugs, and parasitism) may influence BMR; nevertheless, under the simple physiological condition defined above, BMR is so predictable that any significant deviation from normal values for an individual can be used diagnostically. That is one reason why it is held that BMR assesses the energy requirements for maintenance of tissues and organs (plus any dissipation of energy not coupled to ATP formation).

Standard Metabolic Rate

The operational definition of BMR for human beings is usually not applicable to other animals, and especially not to ectotherms. Instead, other sets of operations are utilized (see, for example, Brett and Groves, 1979) to obtain the lowest metabolic rate under a specified set of conditions. This value is called the standard metabolic rate (SMR). Applied to man, the SMR is a perfectly good estimate of human BMR.

In interspecies comparisons, it is well known that the metabolic rate per gram decreases with body weight (M): SMR = aM^b (where a and b are constants). Log–log plots of weight-specific SMR versus M are linear, with a negative slope of about -0.25. SMR values so plotted for most mammals, including man, fall on the same line (for literature in this area, see Calder, 1981; Schmidt-Nielsen, 1984). Other phylogenetic groups display lower or higher absolute SMR values (different values for a), but the values for the exponent b and the slope of log–log plots of SMR versus M are similar to those found for mammals. The basis for the scaling of metabolic rate is discussed by numerous authors elsewhere. Here we merely want to point out that SMR is scaled to body size in a systematic manner. The value of a ($= SMR/M^{0.75}$) for all mammals, for example, is very similar; although it differs from that of other groups such as marsupials or birds, the values for a within each group are very similar. The exponent b, on the other hand, seems to be close to 0.75 for all endotherms, irrespective of phylogenetic relatedness.

Implications of the SMR Concept

Two important consequences stem from the above observations. First, because for any taxon SMR scales to body mass in a similar way, it also follows that for that particular kind of biological organization about the same fraction of potentially available energy must be devoted to SMR. Second, it is necessary that the SMR concept extend down to the tissue and cell level; that is, the SMR for the whole organism equals the sum of the in vivo SMRs of all constituent tissues. Although this has rarely been demonstrated empirically (a technically difficult assignment), it is easy to show that not all tissues contribute equally to SMR (Table 1.2). Put another way, per gram of tissue, SMR is very cell-line specific. The SMR of a gram of nervous tissue, for example, is significantly higher than the SMR of skeletal muscle of the same mass (Table 1.2). Why should this be so? A quiescent cell is a quiescent cell. So should we not expect the energy costs of maintenance to be approximately the same for all cells? To answer this question, we must clarify why an animal at rest requires any energy turnover at all. No nonliving analogue of an adult organism of constant weight and fixed body composition in an isothermal environment would require a source of energy (apart from that required to do work on its environment). Why then should a nongrowing, adult organism at rest in an isothermal environment turn over ATP? In a general sense the answer must be that some ATP-requiring processes normally cannot be fully turned off, even at rest. Identifying what these processes are in effect defines the basis for SMR.

Theoretical Basis for SMR

It is possible to break up the minimal energy requirements of an organism into two classes, depending upon whether or not work is involved. Some organs, such as the heart, must continue to do work on their environment at all times. Otherwise, the organism would die. Thus at least one part of the SMR for the heart represents the energy costs of its work rates under postabsorptive, resting conditions of the whole organism. Identical considerations apply to postural muscles, respiratory muscles, pumps, bailers, cilia, and other continuously working tissues. Within any given organism, their minimal work rates contribute to maintenance metabolic rate.

A second kind of function contributing to minimal energy requirements of organisms is less widely appreciated. Yet every tissue in the body needs to expend considerable amounts of metabolic energy simply to sustain it in a state remote from equilibrium with its environment. Several kinds of equilibrium states must be avoided, the simplest probably being that between cellular macromolecules and their precursors. All cells contain large quantities of polysaccharides, proteins, lipids, and nucleic acids but relatively low concentrations of their constituent precursors. Yet in the presence of appropriate en-

Table 1.2 Relative oxygen consumption rates of different tissues in man

Tissue	At rest[a]	Heavy work
Skeletal muscles	0.30	6.95
Abdominal organs	0.25	0.24
Kidneys	0.07	0.07
Brain	0.20	0.20
Skin	0.02	0.08
Heart	0.11	0.40
Other	0.05	0.06
Sum	1.00	8.00

Source: Modified from McGilvery (1979).

a. The value for the whole body at rest is set at 1.00; the actual $\dot{V}_{O_2}$ at rest is about 0.17 mmol min^{-1} kg^{-1}.

zymes, quite the opposite situation would prevail at equilibrium, and the reaction

$$\text{polymer} \underset{\text{enzymes}}{\overset{\text{hydrolytic}}{\rightleftharpoons}} \text{precursor}$$

would be "pushed" far to the right. Metabolic energy therefore must be expended to synthesize complex cellular constituents such as proteins at a rate equal to that at which they are degraded as the above system tends toward equilibrium.

Organisms also must avoid an equilibrium between the ionic composition and concentration of inner and outer environments. In essentially all organisms, the ionic composition of the extracellular fluid (ECF) bathing body cells is remarkably different from that within cells (intracellular fluid, or ICF), despite permeability of the cell membrane to ions and water from both sides. This disequilibrium depends upon a balancing of two classes of ionic fluxes. The passive leakage of ions through ion-specific, transmembrane proteins called channels is a type of ionic flux that depends strictly on ion-activity (or concentration) differences between the ECF and the ICF. A second type of flux is an active, ATP-requiring transport of ions via ion-specific, transmembrane proteins called ion pumps; as the name implies, this flux can operate against (sometimes very large) diffusion gradients. The maintenance of disequilibria thus depends upon balancing ionic leakage rates and ionic pumping rates (upon balancing fluxes through ion-specific channels and ion-specific pumps) at specific set points or transmembrane electrochemical potentials and are only sustainable through the continuous expenditure of metabolic energy.

In all such instances of disequilibrium, thermodynamically unfavorable processes in effect need to be facilitated, and this is most commonly achieved by coupling them to ATP hydrolysis (Scheme 1.1). It is important to emphasize that in each case reaction I is not the

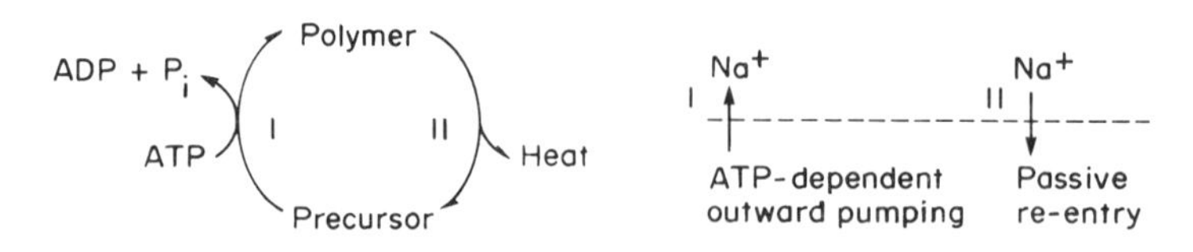

Scheme 1.1

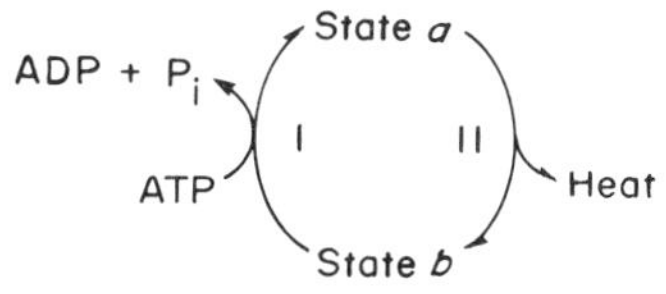

ADP + P_i ... State a ... I ... II ... Heat ... ATP ... State b

Scheme 1.2

reverse of reaction II. The two arrows represent spatially distinct pathways that are theoretically reversible, but that under physiological conditions can usually be considered as effectively unidirectional. The spatial separation of the two analogous processes in Na^+ exchange across cell membranes, as illustrated in Scheme 1.1, makes the point most clearly: metabolic energy (ATP) is expended in path I to keep path II displaced from its inherent equilibrium. Scheme 1.2 illustrates the general case, where state *a* represents chemical, osmotic, or physical conditions displaced from equilibrium for reaction path II and state *b* represents the equilibrium for reaction II, which in reality often means chemical, osmotic, or thermal equilibrium of the organism with its environment. ATP must be expended in reaction path I at the same rate as reaction path II tends toward state *b* (and thus toward equilibrium). The net reaction is the hydrolysis of ATP with a consequent release of heat. The sum of all such processes in the organism may be presumed to constitute a fraction of the SMR—but is it a large fraction? Although we cannot answer the question for many (perhaps most) such functions, we can with confidence estimate these costs for at least two: protein turnover and Na^+,K^+ pumping. Except under conditions of rapid growth, the energy cost of protein turnover is a modest fraction of the SMR, probably about 5% or less. In some cells (reticulocytes, for example), however, protein biosynthesis may account for up to 30% of the basal rates of ATP turnover (Rapoport, 1985). By comparison, the metabolic costs of one transport function alone—that of maintaining low Na^+ and high K^+ concentrations in the cell—ranges between 30 and 60% of the SMR (Hulbert and Else, 1981). In reticulocytes, it accounts for 25% of the cell SMR. Because active transport of other ions and metabolites is the rule rather than the exception, we feel it is safe to conclude that a large fraction of the SMR represents energy expenditure for keeping tissues and organs remote from ionic equilibrium with their environment.

Metabolic Arrest and the Time Extension Factor

The preceding analysis implies that metabolic rate can be reduced to a level below the SMR level by two primary and necessarily coupled processes: (1) suppression of oxidative and anaerobic pathways of ATP production; and (2) proportionate reduction of the rates of various ATP-requiring "maintenance" functions, in particular by proportionately reducing the permeability of barriers between cells, tissues, organs, or even organisms and their environments. Although during modest degrees of metabolic suppression these two processes may predominate, in extremely dormant states (in cold torpor, in complete freezing, or in complete dehydration), intracellular conditions may be drastically altered. Under the new conditions, intracellular homeostasis, which is normally sustained by maintenance metabolism, cannot be maintained; and this breakdown may be viewed as one cost of being able to enter deeply arrested states. Sustained survival under these conditions may be assured only by linking these two processes with a third: (3) harnessing of mechanisms for protecting intracellular structures against damage or denaturation stemming from the altered intracellular milieu. Whereas these mechanisms are first and foremost protective measures against environmental hazards, they also usually serve as negative feedback loops and even further limit normal cell functions, because "protected" intracellular structures (enzymes, mRNA, ribosomes, and so on) are also usually nonfunctional. In fact, we might expect that the deeper the dormancy, the greater the role of provision (3) in achieving the arrested state. Provision (3) in this view becomes a means for amplifying the effectiveness of provisions (1) and (2).

Switching down metabolism well below SMR levels thus seems readily achievable, at least in theory. A consequence of using such mechanisms under conditions of environmental stress is that at reduced metabolic rates any given biological process (whether it be a systole–diastole cycle, a substrate interconversion cycle, or a catalytic cycle) is proportionately slowed down. Lindstedt and Calder (1981) especially have emphasized the generality of this relationship between metabolic rate and biological time: at all levels of biological organization, a fractional change in metabolic rate means a fractional change in biological time (relative to astronomical time) of equal magnitude but opposite sign (Table 1.1). The time extension factor is the degree to which organisms can utilize these strategies or the degree to which they can arrest their metabolism and thus extend time.

How useful is this ability and how widely is it used? We will begin our analysis of these questions by turning our attention to facultatively anaerobic animals that utilize a modest degree of metabolic arrest as their most effective defense against hypoxia. By slowing down metabolism and extending time (a time extension factor of 5–20), they are able to sustain degrees of hypoxic stress that would otherwise be debilitating and probably lethal.

2

Animal Anaerobes

Although most animals, including man, are extremely sensitive to even modest O_2 limitations, numerous species are known to be profoundly resistant to hypoxia. A close analysis of such animals indicates that they typically rely on only one of two broad categories of adaptive response. In the first type of response, metabolic strategies are directed toward *sustained oxidative function,* despite potentially chronic O_2 limitations. This adaptive response, in which there is little or no extension in tissue anaerobic capacities, is found in high altitude-adapted animal species and in human beings suffering from chronic hypoxia. In contrast, numerous invertebrates and lower vertebrates under hypoxic stress rely critically on metabolic strategies directed toward *sustained anaerobic function.* Many of these species have developed such effective mechanisms of protection against hypoxia that they are often referred to as "good" animal anaerobes, or more precisely as facultative anaerobes (Hochachka, 1985, 1986). They often display two important responses to varying O_2 availability that are generally overlooked. One of these responses is O_2 conformity: as O_2 availability in the inspired medium declines, the rate of O_2 uptake ($\dot{V}_{O_2}$) also declines. A second, perhaps more remarkable, response is electron transfer system (ETS) arrest: as O_2 availability declines to some critical level, $\dot{V}_{O_2}$ rather abruptly stops.

These responses raise three questions that apply broadly to all systems showing similar metabolic behavior. First, how is oxidative metabolism suppressed and, at the limit, arrested as O_2 availability declines? Second, how does the organism avoid an energetic shortfall? And third, how are cell maintenance functions balanced with the reduced rates of ATP synthesis? We shall consider these issues in turn.

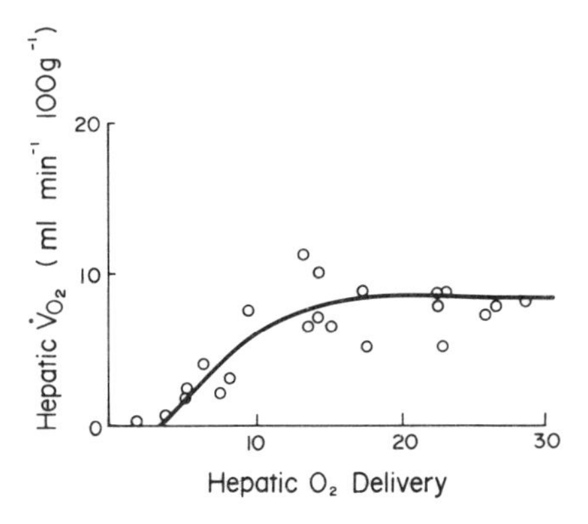

Figure 2.1 Pattern of O_2-conforming $\dot{V}_{O_2}$ responses to O_2 availability in mammalian liver, indicating a case in which ETS arrest occurs before O_2 supplies become fully depleted. Modified from Edelstone et al. (1984).

O_2 Conformity: Metabolic Suppression through O_2 Limitation

Cells, tissues, or organisms whose $\dot{V}_{O_2}$ varies with O_2 availability over broad ranges are termed O_2 conformers to highlight the contrast between their behavior and the behavior of O_2 regulators, whose $\dot{V}_{O_2}$ is independent of O_2 availability down to very low values. The metabolic response of the mammalian brain to changes in O_2 availability fits the pattern of O_2 regulators (Kinter et al., 1984) and represents an extreme end of a spectrum of responses observable in mammalian tissues. Liver expresses an intermediate pattern (Figure 2.1); and skeletal muscle is perhaps the most O_2-conforming of all mammalian tissues. At least in some mammals, no plateau in muscle O_2 consumption is reached, even at very high O_2 tensions or very high O_2 delivery rates (Figure 2.2). Similar O_2-conforming patterns are common among numerous invertebrate groups and among ectothermic vertebrates (see Mangum and Van Winkle, 1973).

In contrast, isolated mitochondria are universally found to display very high O_2 affinities and respiration rates that are largely independent of O_2 concentration. Their respiratory patterns are clearly those of O_2 regulators—as they should be, because the K_m for O_2 is usually in the 10^{-6} to 10^{-7} M range (Figure 2.3). Thus O_2-conforming patterns displayed by a wide diversity of organisms present us with an interesting metabolic paradox: Why should $\dot{V}_{O_2}$ be declining at O_2 concentrations that are fully saturating for isolated mitochondria? Our analy-

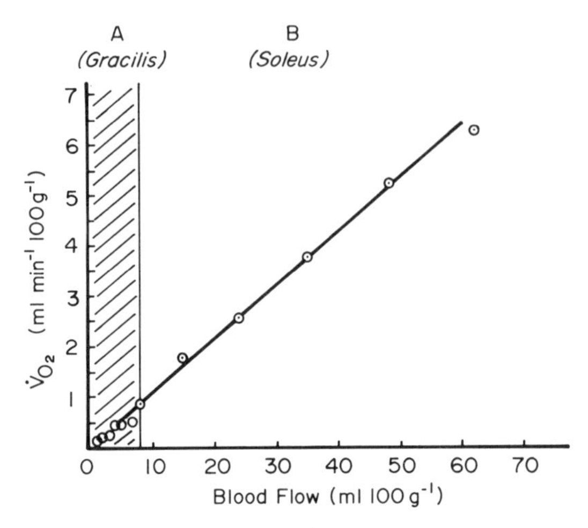

Figure 2.2 The relationship between $\dot{V}_{O_2}$ for muscle and O_2 availability (represented by blood flow) for grouped data from nine experiments on gracilis muscle (A) and ten on soleus (B) of the cat. Each point represents weighted mean of values from three or more experiments. Modified from Whalen et al. (1973).

sis of this problem indicates that there are three current ways to look at the problem (see Appendix A).

One possible explanation assumes O_2 sensing. In this model, O_2 concentration serves as signal for a respiratory set point; but how an O_2 receptor may work, if it exists, is unknown. A second possibility is

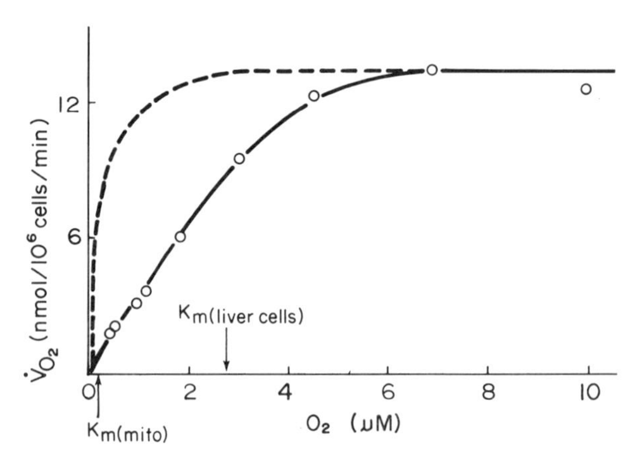

Figure 2.3 O_2-conforming (*solid*) versus O_2-regulating (*dashed*) responses of $\dot{V}_{O_2}$ as a function of O_2 concentrations. Modified from Jones and Kennedy (1982).

that the electron transfer system per se involves a reversible O_2 binding step which is close to equilibrium with all earlier steps in the chain. If so, this could explain how mitochondrial O_2 uptake rates can respond sensitively to O_2 concentration. Whereas this model is mechanistic and heuristic, it is not at this point widely accepted, partly because cytochrome oxidase catalytic mechanisms are still poorly understood. A third alternative simply assumes true O_2 limitation under in vivo conditions due to O_2 sinks coupled with diffusion barriers.

Evidence in favor of this interpretation arises from comparisons of the O_2 dependence of different intracellular O_2-requiring functions, such as O_2 binding by myoglobin and urate oxidase and reduction of ETS components. As O_2 availability declines, all three functions decline in parallel in vivo despite very different in vitro O_2 dependencies (very different in vitro O_2 affinities). Over a decade ago, Chance (1976) took such data to indicate very steep $[O_2]$ gradients, with $[O_2]$ at the mitochondria being much lower than would be anticipated from $[O_2]$ in plasma; that seems still to be a favored interpretation today (Chance and Leigh, 1985; Jones et al., 1985). True O_2 limitation of mitochondrial oxidative metabolism as O_2 availability declines seems the only possible explanation for O_2 conformity in the lungless salamander (Gatz and Piiper, 1979) and is the alternative we would favor for O_2 conformity in facultative anaerobes because it is consistent with a limited O_2 delivery capacity—limited by capacities of ventilation, cardiac output, and blood-to-tissue transport (Johansen et al., 1978; Lykkeboe and Johansen, 1982). In fact, potentially limited O_2-delivery capacities are typically found in mammalian O_2-conforming tissues as well, so the concept that plots of $\dot{V}_{O_2}$ versus $[O_2]$, although right-shifted on the $[O_2]$-axis, nevertheless reflect the form of true O_2 saturation curves for working mitochondria, may be universally applicable. However, even if this interpretation can adequately explain O_2 conformity down to low P_{O_2}, it does little to account for $\dot{V}_{O_2}$ being fully arrested at this O_2 availability. A new and different mechanism must be set in motion at this point.

Arresting Oxidative Metabolism

Whereas O_2 limitation may be the most probable explanation for O_2 conformity, the most plausible of potential mechanisms of ETS arrest appears to involve adenylate translocation. It is known that, in a variety of systems during normoxia–hypoxia transitions, the steady-state distribution of adenylates is shifted in favor of the cytosolic pool (Aprille and Brennan, 1985; Austin and Aprille, 1984; Brennan and Aprille, 1985); there is no reason to believe that the same process does not occur generally in O_2-conforming animals as O_2 availability declines. In effect, the steady-state distribution of adenylates

$$\text{cytosolic adenylates} \underset{\text{normoxia}}{\overset{\text{hypoxia}}{\rightleftharpoons}} \text{mitochondrial adenylates} \quad (1)$$

is shifted to the left during hypoxia, and various ADP- and ATP-linked mitochondrial processes, including the synthesis rates of urea, glucose, and ATP, decline accordingly. These shifts in adenylate distribution, presumably coupled with O_2 limitation in all forms of O_2-conforming behavior, are fully and rapidly reversible in vitro (Tagawa et al., 1985) and in vivo (Brennan and Aprille, 1985). We suggest that O_2 conformity becomes ETS arrest when adenylate translocation to the cytosol is complete, for at that point most, and probably all, mitochondrial functions linked to adenylate metabolism become necessarily, if reversibly, arrested.

In this view, then, ETS arrest is O_2 conformity carried to completion (mitochondrial adenylate pools fully depleted), an interpretation that is helpful in explaining why this kind of behavior is most commonly observed in animals well adapted to hypoxia. The tissues in these organisms are poor in mitochondria and the bulk of the adenylate pool is already cytosolic even in normoxia. In such systems, O_2-conforming responses to declining $[O_2]$ presumably would more rapidly deplete mitochondrial adenylate pools than in homologous, mitochondria-rich tissues. Thus ETS arrest (that is, complete depletion of mitochondrial adenylates) frequently may be expected to cut in before O_2 is fully depleted (see Figure 2.1). Indeed, this may occur at different $[O_2]$ in different tissues and in different species, a prediction consistent with observations (Mangum and Van Winkle, 1973). In some systems, of course, the process may not reach completion until very low $[O_2]$ (O_2 conformity without ETS arrest)—also a situation commonly observed. For several reasons, then, the adenylate translocation hypothesis (Aprille and Brennan, 1985) seems particularly useful in explaining the various characteristics of ETS arrest and its relationship to O_2 conformity.

Metabolic Mechanisms Protecting "Good" Animal Anaerobes against Anoxia

The next problem is how the organism gets by when oxidative metabolism is fully blocked. In considering this problem, we have used phylogenetically diverse groups of facultative anaerobes to assess the most effective metabolic mechanisms utilized in the protection of tissues and organs against O_2 lack. All such organisms behave as closed or semi-isolated, anaerobic life support systems faced with two critical problems: (1) conservation of fermentable substrate, and (2) avoidance of self-pollution by accumulation of undesirable end products. The first problem arises from the energetic inefficiency of anaerobic metabolism, for no matter which substrates are utilized (carbohydrates or amino acids; and in mammals, the former predominate), the yield of ATP per mole of substrate fermented is always modest compared with the ATP yield of oxidative metabolism. For this reason in most animal tissues carbohydrate consumption rates are inversely related to O_2 availability, which means that, if demands for ATP remain unchanged during anoxia, these rates necessarily have to rise drastically, in a process termed the Pasteur effect (see Appendix B).

The consequent problem of potentially massive depletions of glycogen from central stores is minimized in "good" animal anaerobes either by (1) storing more glycogen in peripheral depots, (2) utilizing more efficient fermentation pathways, or (3) depressing ATP synthesis rates during O_2 limiting periods, or some combination of (1), (2), and (3). All three mechanisms are demonstrably useful by the criterion of anoxic survival time. In "good" animal anaerobes, such as *Mytilus*, goldfish, or diving turtles, tissue glycogen is stored at concentrations that are as much as three to four times higher than in other species; this strategy alone, therefore, could extend hypoxia tolerance by the same (3- to 4-fold) factor. Similarly, the most energetically efficient fermentation pathway known in animal tissues (glycogen → propionate) achieves about 3-fold more ATP per glucosyl unit than does classic glycolysis (Figure 2.4), and thus confers an advantage of the same magnitude. Compared with these maximally 3- to 4-fold effects, metabolic arrest processes in goldfish can increase anoxia tolerance by about 5-fold; in *Mytilus*, by about 20-fold; and in brine shrimp embryos, by several orders of magnitude. Brine shrimp embryos in estivation actually illustrate this strategy at its limit, for their maximum anoxia tolerance coincides with entrance into a fully arrested or ametabolic state (see Chapter 9). Nevertheless, the exam-

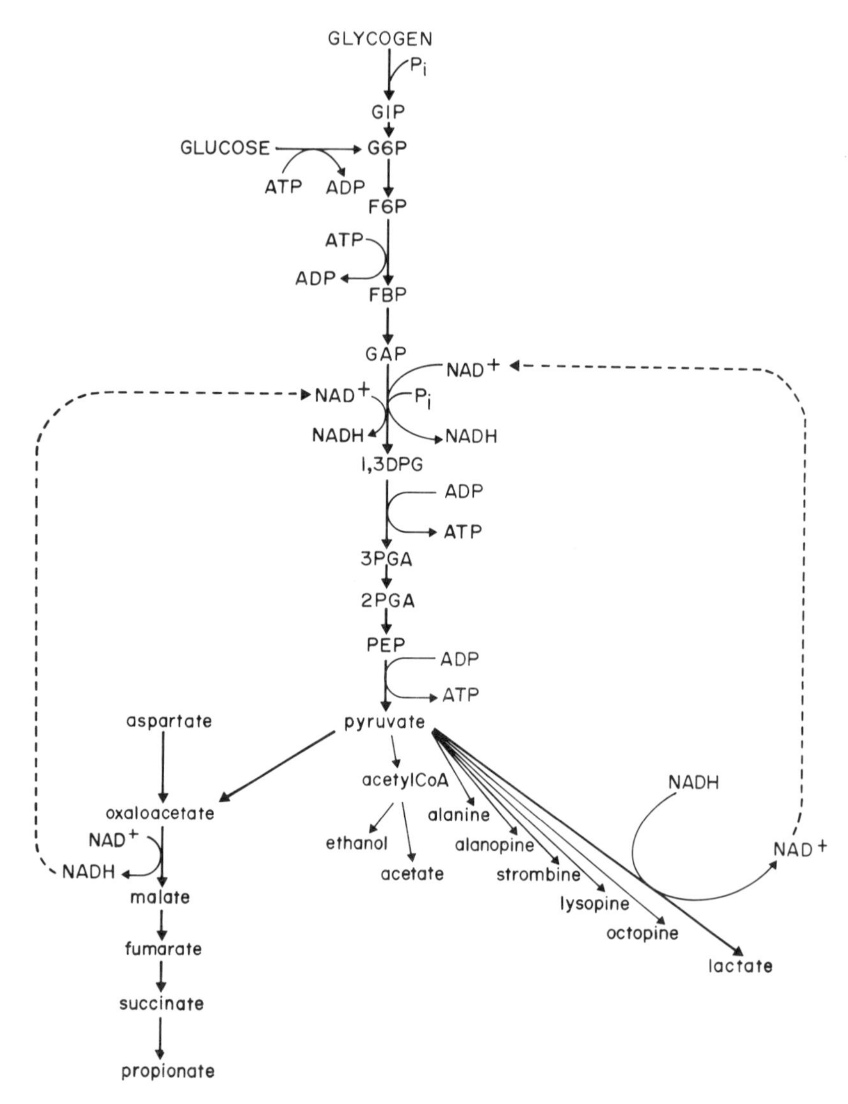

Figure 2.4 O_2-limited metabolism in facultatively anaerobic animals. Current concepts view the organization of anaerobic metabolism in these animals as a series of linear, and loosely linked, pathways (see Table 2.1). Pathways 2–5 are known in various bivalve mollusks: 3, 4, 9, and 10 are often used by helminths; 6–8, while theoretically possible in bivalve mollusks, do not appear to be used to any significant extent. Pathway 11, or sulfate oxidation of organic substrates, is well known to occur in the highly reduced layers of benthic silt, but the distribution of this activity among bacteria and lower invertebrate animals is not yet clarified. Pathways 8–11 are not shown in the figure. From Hochachka and Somero (1984); reprinted by permission of Princeton University Press.

ples available from less extreme animal anaerobes lead to two instructive insights. First, reversing the classic Pasteur effect to allow ATP turnover rates to drastically decrease during anoxia appears to constitute the most effective strategy for solving the first of the aforementioned problems (substrate conservation). Second, and perhaps equally important, theoretically this metabolic strategy is universally available, whereas alternative mechanisms (such as fermentation pathways with improved ATP yield) have limited phylogenetic distributions and, therefore, even in theory could not represent realizable strategies of anoxia adaptation in species lacking the appropriate enzyme pathways—for example, most vertebrates. So our estimates of the functional advantages of metabolic arrest processes for surviving periods of O_2 lack may actually be on the conservative side—assuming, of course, that all else is equal. Obviously it never is, which brings us to the second problem of end product accumulation.

Our first analysis of the "end products problem" was necessarily more complex because it required an assessment of the relative effects of organic (usually anionic) end products and those of H^+. The metabolic sites of H^+ production and the pathways for subsequent proton deposition had to be considered, as did the H^+ stoichiometry of dif-

Table 2.1 Pathways of anaerobic metabolism in facultatively anaerobic animals

Pathway	Substrate		Product	Energy yield (moles ATP per mole of substrate)
1	Glucose	→	Lactate	2
2	Glucose	→	Octopine, lysopine, alanopine, or strombine	2
3	Glucose	→	Succinate	4
4	Glucose	→	Propionate	6
5	Glucose	→	Acetate	4
6	Aspartate	→	Succinate	1
7	Aspartate	→	Propionate	2
8	Glutamate	→	Succinate	1
9	Glutamate	→	Propionate	2
10	Branched-chain amino acids	→	Volatile fatty acids	1
11	$4CH_2O + 2SO_4^{--} + 3H^+$	→	$H_2S + HS^- + 4H_2O + 4CO_2$	6

ferent fermentation pathways. Despite these difficulties, it became evident that the potentially perturbing effects of H^+ and organic anion accumulations are minimized in "good" animal anaerobes by only a handful of mechanisms. These include (1) utilizing fermentation pathways that allow more ATP to be turned over per mole of H^+ accumulated than in classic glycolysis, (2) tolerating proton accumulation by improved tissue buffering capacity, (3) minimizing end product accumulation by recycling it for further metabolism or excretion, (4) utilizing H^+-consuming reaction pathways, and (5) depressing metabolic rates during anoxia.

Of these, the last is again by far the most effective, a conclusion that can be illustrated by considering each mechanism in turn. For mechanism 1, a maximum of 3 mol H^+/mol ATP cycled is theoretically obtained by coupling the glucose → propionate pathway with cell ATPases, compared with 1 mol H^+/mol ATP when the glucose → lactate pathway is coupled to ATPases (Table 2.2). For mechanism 2, buffering capacity is typically high in glycolytic tissues and lower in oxidative ones; and there is an interspecies relationship between buffering and anaerobic capacities. Between-species comparisons among vertebrates indicate a 3-fold range of muscle buffering values (about 40–110 μmol/g). Among invertebrates the absolute values are lower (20–60 μmol/g), but the range over which adaptational adjustments seem possible again is only about 3-fold. In man the range may be even less. The buffering capacities of homologous muscle from ath-

Table 2.2 Relationship between ATP cycled (between fermentation pathway and ATPases) and H^+ produced

Pathway coupled with ATPases[a]		ATP cycled/H^+ produced
glucose	→ lactate	1.0
glycogen	→ lactate	1.5
	→ alanopine	
	→ strombine	
	→ octopine	
glucose	→ succinate	2.0
glucose	→ propionate	3.0

Source: After Hochachka and Mommsen (1983).

a. The advantage to "good" anaerobes of using alternate fermentations is that more ATP can be turned over per H^+ produced.

letes trained for anaerobic events are about 20 μmol/g higher than the buffering capacities in athletes trained for aerobic work. If we assume maximum lactate levels to be about 40 μmol/g muscle, this adjustment could allow an increase in H^+ production of about 1.5-fold. Thus for a variety of species, mechanism 2 may be expected to yield a 1.5- to 3-fold advantage in tolerating end product (H^+) accumulation. This estimate probably represents a maximum, because such adjustments may not be possible in tissues other than muscle.

The effectiveness of mechanism 3 is difficult to assess. Recycling of lactate formed in hypoperfused peripheral tissues for oxidation in better-perfused organs apparently occurs during diving in the Weddell seal, but it accounts for only a modest fraction of the total lactate and protons accumulated during the experimental period. Because the Weddell seal may have a high hypoxia tolerance, possibly required during maximum duration (1.2-hr) breath-hold diving, we tentatively conclude that this mechanism for minimizing end products problems is not quantitatively significant here or elsewhere. However, it is a mechanism that begs for further investigation.

The maximum effectiveness of mechanism 4, using H^+-consuming reaction sequences such as

$$\text{AMP} \rightarrow \text{IMP} + \text{NH}_3 \xrightarrow{\text{H}^+} \text{NH}_4{}^+$$

can be easily illustrated by assuming that essentially all cell ATP could be converted to AMP, which on deamination (catalyzed by AMP deaminase) would release NH_3. For most cells, a maximum of about 5 μmol/g of H^+ could be utilized to protonate the ammonia, which is, of course, a modest fraction of the total proton and lactate loads that are observed during O_2 lack.

Whereas reliance on any one of mechanisms 1–4 is obviously advantageous, yielding up to a several-fold improvement in tolerance of O_2 lack, it is evident that by depressing demands for ATP during anoxia (mechanism 5), an organism not only reduces the depletion rates of carbohydrate in inefficient fermentations, it also automatically reduces rates of formation of anaerobic end products, including H^+. In the anoxic goldfish the rate of proton production is reduced 5-fold as a result of metabolic depression, whereas in the turtle it is reduced approximately 60-fold. This may be particularly advantageous if proton efflux can be disassociated from that of lactate, when proton efflux from tissues, even if it occurs at low rates, may be able to pace rates of proton production in tissues.

The picture emerging from such analyses of phylogenetically diverse groups of animals is that several processes contribute to the anoxia tolerance of "good" anaerobes, but of these, the arrest strategy extended to anaerobic metabolism yields by far the most effective protection against O_2 lack. Of the known protective strategies, it alone supplies resolution to both the problem of substrate conservation during anoxia and the problem of end product self-pollution. That is why it has become evident in recent years that for prolonged survival in hypoxia or anoxia, anaerobic life-support systems (whether considered at the organismal, organ, or cellular level) must be able to switch-down anaerobic ATP synthesis rates even more than observed in oxidative arrest.

We will call this model the glycolytic arrest concept of defense against hypoxia, although it is more accurately described as an arrest of the glycolytic activation that is normally expected to make up energetic shortfalls arising from O_2 lack. This feature is of obvious advantage in natural systems.

Mechanism of Glycolytic Arrest during Oxygen Limitation

In principle, glycolytic arrest could be achieved either by slowing down the rate of ATP synthesis or by slowing down the rate of ATP utilization during anoxia (Scheme 2.1). Most animal anaerobes seem to choose the former (the left arm of the ATP–ADP cycle) and are faced with a two-pronged problem: how to avoid activation of a Pasteur effect, and, in the extreme, how to actually reverse it.

Hypoxia-tolerant systems commonly display O_2 conformity. Once such a system is under complete anoxia, the mitochondrial pool may become almost totally depleted, a condition leading to increased Ca^{++} and ADP availability, both of which would favor glycolytic activation (Tagawa et al., 1985). For this reason, and because other metabolite

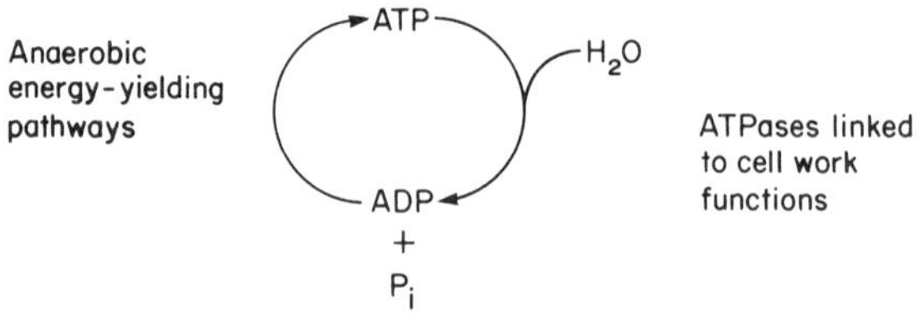

Anaerobic energy-yielding pathways

ATP

H_2O

ATPases linked to cell work functions

ADP + P_i

Scheme 2.1

changes at key loci in glycolysis (such as positive modulators for phosphofructokinase) all are consistent with glycolytic activation, we would anticipate a powerful Pasteur effect. Indeed, this does occur in many, if not most, O_2 regulators (and it is one of the processes leading to cell damage and one of the factors causing these animals to be hypoxia-sensitive). In contrast, hypoxia-tolerant animals behave as if mechanisms are activated for overriding these initial signals for a Pasteur effect. At their most effective, these override effects can actually *reduce* glycolytic fluxes and ATP turnover rates to values lower than those in normoxic or hypoxic states, despite O_2 lack. While the mystery of the missing Pasteur effect is not yet resolved (see Appendix B), it is already evident that it could be modified in a number of ways (Storey, 1985). These include (1) modifications in allosteric regulation of key glycolytic enzymes, (2) covalent modification of key regulatory enzymes, and (3) modification of the three-dimensional pathway structure and function via cell-line-specific adjustment in controlling processes dependent upon enzyme–enzyme, enzyme–myofibrillar, or enzyme–band 3 protein interactions (see Appendix B); actual mechanisms used in any given system may well be cell-line-specific and even pathway specific. In functional terms, perhaps the most significant component of glycolytic arrest, however, is that of biological time extension, as seen in O_2 conformity and in ETS arrest.

In summary, then, we are faced with an interesting situation in which numerous hypoxia-tolerant systems have three potential, short-term mechanisms for reducing rates of ATP synthesis when O_2 availability is limiting (Table 2.3). The first of these is true O_2 limitation leading to O_2 conformity and partial depletion of mitochondrial adenylates. The second is ETS arrest, which may be expected to occur

Table 2.3 Short-term mechanisms of metabolic arrest in hypoxia-tolerant organisms

Process	Possible mechanisms
O_2 conformity	O_2 limitation (partial loss of mitochondrial adenylates)
ETS arrest	Complete O_2 depletion, or complete adenylate depletion, or both
Glycolytic arrest	Reversed Pasteur effect

Note: Long-term mechanisms of arresting cell function include biosynthetic blockade (translational blocks, nucleolar segregation, inhibition of protein synthesis, and even cell division blockade). These mechanisms will be discussed in Chapters 9 and 10.

either when mitochondrial adenylates are critically depleted or when $[O_2]$ drops to vanishingly low values, prohibitive of ETS function. And finally, when the system is fully anoxic, glycolytic arrest (the reversed Pasteur effect) assures that the strategy of switching off in the face of O_2 lack is extended to anaerobic ATP-generating pathways. In earlier discussions of hypoxia tolerance, the combination of these processes is termed the metabolic arrest concept of defense against hypoxia (Hochachka, 1982, 1986; Hochachka and Dunn, 1983), and it interesting to note that it has an analogue in clinical studies.

Metabolic Arrest as an Intervention Strategy: Mammalian Test Cases

Analogous "arrest"-type concepts of protection against hypoxia are evident in the scientific literature on hibernation and in the clinical literature on cardiac arrest, stroke, acute renal failure, and liver ischemia. It is widely accepted, for example, that ischemic myocardial damage is a function of work load or metabolic rate. Several intervention procedures attempt to minimize tissue damage by reducing myocardial energy requirements during ischemia; and, to a lesser or greater degree, all such maneuvers are helpful—and so consistent with the strategy utilized by "good" animal anaerobes. However, the most convincing test of our metabolic arrest hypothesis is provided by recent studies using ischemic rat kidney preparations as models of acute renal failure. During ischemic acute renal failure, reductions in renal blood flow occur with consequent reduction in delivery of O_2 and substrates to the tissue. Because the main ATP-requiring processes of the kidney involve membrane-coupled ion translocations, our hypothesis leads to an explicit, testable prediction—namely, that experimentally reducing the demands for ATP by ion pumps should yield a proportionate increase in tolerance to O_2 lack. In mammalian kidneys the medullary thick ascending limb (mTAL) of Henle's loop is the most hypoxia-sensitive segment of the nephron; and during perfusion of the isolated organ extensive, essentially irreversible, damage to mTAL cells occurs in 90 min. But this hypoxia sensitivity can be fully eliminated by perfusion with ouabain, a specific inhibitor of Na^+,K^+-ATPase, or by reducing ion pumping work by preventing glomerular filtration. Conversely, polyene antibiotics increase membrane permeability, increase the energy requirements of ion transport, and consequently increase the hypoxia sensitivity of mTAL tubules; yet again the hypoxia-induced damage is prevented if reabsorptive transport is inhibited with ouabain. Full protection can be achieved even against KCN-induced lesions by simultaneous ouabain inhibition of Na^+,K^+-ATPase (Brezis et al., 1984).

These foresightful studies effectively represent experimental attempts to demonstrate that a mammalian organ, in terms of hypoxia tolerance, can be "made" to become similar to hypoxia-adapted lower vertebrates. Yet they also emphasize a critical difference; namely, that in metabolically arrested mammalian systems, hypoxic survival times are only extended from minutes to hours, whereas "good" animal anaerobes are able to sustain anoxia for days, weeks, and even months. Such order-of-magnitude differences in hypoxia tolerance between mammalian preparations and facultatively anaerobic animals means that something is still missing in our analysis of strategies for protecting tissues against O_2 lack. The missing element probably is to be found at the interface between cell membrane and cell metabolic functions and in the way maintenance requirements are satisfied as ATP synthesis rates decline in anoxia.

Balancing Metabolic and Membrane Functions

To illustrate the fundamental importance of close coupling between cell metabolism and cell membrane processes, there is probably no more convenient a tissue or organ than the mammalian brain, which is one of the most hypoxia-sensitive parts of the mammalian body and for which events during various kinds of energy perturbations are well charted and reviewed (Siesjo, 1981; Hansen, 1985). The principles that emerge, however, are applicable to all hypoxia-sensitive cells and tissues. In complete cerebral ischemia, then, as occurs in cardiac arrest (Figure 2.5), the EEG becomes isoelectric within 15–25 sec. This electrically silent period precedes a massive outflux of K^+ from the neurons and an influx of Na^+ into the neurons; these ion fluxes are due to energy insufficiency and thus to the failure of membrane ion pumps (at regional cerebral blood flow of about 10% of normal). At an ECF $[K^+]$ of 12–13 m*M*, changes in membrane potential apparently become large enough to activate (or open) voltage-dependent Ca^{++} channels and develop a largely uncontrollable influx of Ca^{++}, a cation that at abnormally high cytosolic concentration acts as a cellular toxin. Although high cytosolic $[Ca^{++}]$ may disrupt vari-

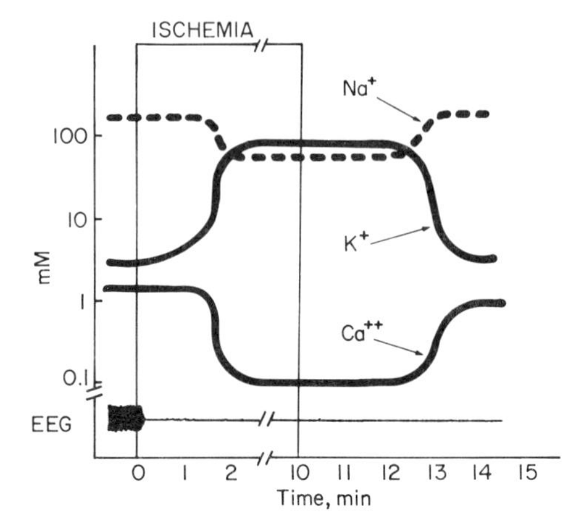

Figure 2.5 Extracellular ion concentration and EEG activity in rat brain cortex at the onset and end of 10 min of cerebral ischemia. From Hochachka (1986); copyright 1986 by the AAAS.

ous intracellular functions, its activation of phospholipases A_1 and A_2 is considered to be the most damaging under hypoxic conditions. Activation of phospholipase A_2 is shown in Scheme 2.2. Uncontrolled, this reaction leads (Figure 2.6) to membrane phospholipid hydrolysis, to the consequent disruption of cell and mitochondrial membranes, to the release of unique free fatty acids (such as arachidonic acid), and to the further potentiation of ion redistributions.

Potentially damaging, ion-flux-initiated perturbations also may be facilitated by the action of phospholipase C, whose continued catalytic function under O_2-limiting conditions (Matthys et al., 1984) is indicated by increasing levels of stearoyl and arachidonoyl diacylgycerols coincident with decreasing levels of phosphatidylinositol (PI). The major metabolic pathways responsible for the formation and degradation of PI are illustrated in Figure 2.7. Phosphatidylinositol is phosphorylated at the 4-position of its inositol head group by a specific

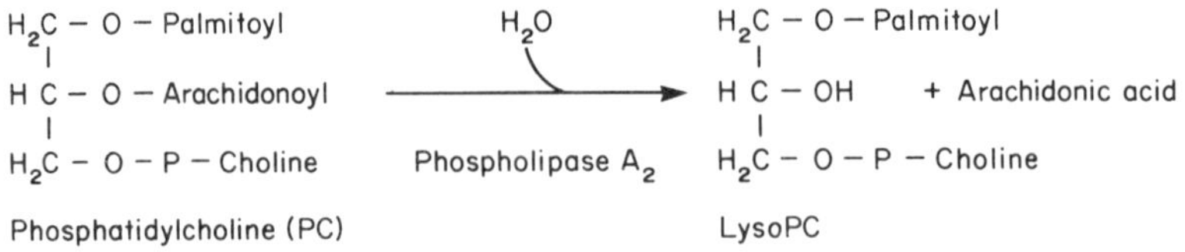

H_2C – O – Palmitoyl | H C – O – Arachidonoyl | H_2C – O – P – Choline
Phosphatidylcholine (PC)

$\xrightarrow[\text{Phospholipase } A_2]{H_2O}$

H_2C – O – Palmitoyl | H C – OH | H_2C – O – P – Choline + Arachidonic acid
LysoPC

Scheme 2.2

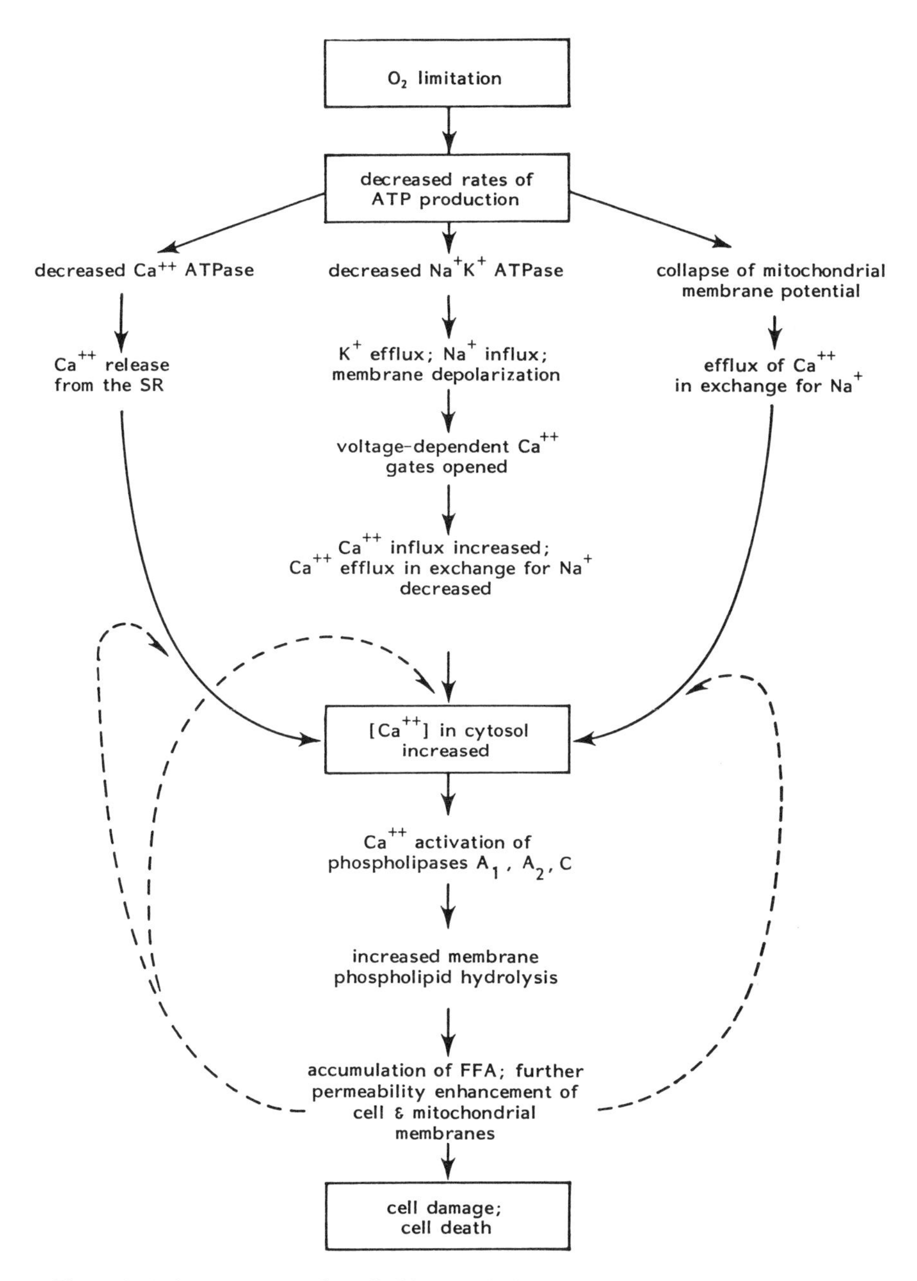

Figure 2.6 A summary of probable metabolic events occurring in hypoxia-sensitive cells and progressing from the initial energetic consequences of O_2 limitation to cell damage and cell death. The summary is based on analysis in the text and is constructed from various studies of hypoxia-sensitive mammalian tissues. SR, sarcoplasmic reticulum; FFA, free fatty acids. From Hochachka (1986); copyright 1986 by the AAAS.

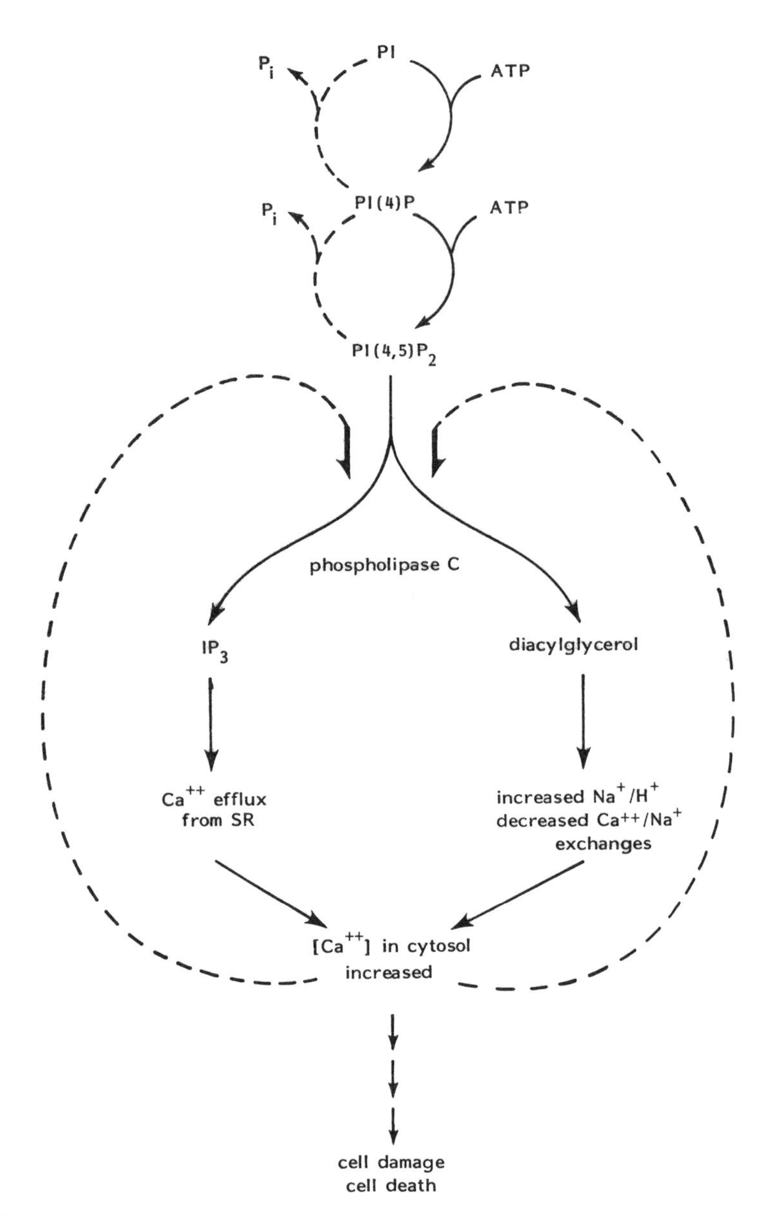

Figure 2.7 Under O_2-limiting conditions, phosphatidylinositol (PI) decreases in concentration and diacylglycerol increases (as described for the ischemic kidney; Matthys et al., 1984). These changes are presumed to occur along with some accumulation of inositol triphosphate (IP_3), further facilitating increased Ca^{++} levels in the cytosol. See Berridge and Irvine (1984) for a discussion of how this normally operates as a closely tuned signal-transduction system controlled by Ca^{++}-mobilizing hormones. From Hochachka (1986); copyright 1986 by the AAAS.

kinase, thereby forming phosphatidylinositol 4-phosphate (PI(4)P); this intermediate is further phosphorylated at the 5-position to give $PI(4,5)P_2$, which is one of the inositol lipids located in the inner leaflet of the plasma membrane. The steady-state concentration of $PI(4,5)P_2$ is determined by the balance between the activities of these kinases and phosphomonoesterases that convert $PI(4,5)P_2$ back to PI (Figure 2.7)—that is, by the operation of two linked metabolic cycles whereby phosphates are constantly being added to and removed from the 4- and 5-positions of the inositol head group (Berridge and Irvine, 1984).

In response to various Ca^{++}-mobilizing hormones, these two metabolic cycles are broken in a controlled way by preferential phospholipase C action upon $PI(4,5)P_2$, thereby releasing diacylglycerol and water-soluble inositol triphosphate (IP_3) as a second messenger signaling release of Ca^{++} from intracellular pools. In hypoxia these cycles are broken in an apparently uncontrolled way, presumably leading to the same end products. The IP_3 released in the process is thought to act as a secondary signal for opening Ca^{++} channels and releasing sarcoplasmic Ca^{++}, thus increasing cytosolic Ca^{++} availability. This cascade also may be self-potentiating (increasing Ca^{++} availability favoring phospholipase C catalysis). Under normal circumstances diacylglycerol is phosphorylated to phosphatidic acid, which is then converted back to PI. But in response to appropriate signals and in hypoxia, diacylglycerol accumulates and is thought to increase Na^+–H^+ exchange, thus effectively slowing down Na^+ exchange-based Ca^{++} efflux (Figure 2.7).

The overall impact of O_2 limitation to hypoxia-sensitive nervous systems, as summarized in Figure 2.6, is similar in other mammalian tissues as well. In myocardial ischemia, energy-deficiencies and membrane failures are indicated by intra- and extracellular changes in $[Na^+]$ and $[K^+]$ as well as by a large influx of Ca^{++}, a loss of sarcolemmal Ca^{++}, and a disruption of mitochondrial Ca^{++} homeostasis (Fleckenstein et al., 1983; Nayler, 1983). Analogous membrane failure with associated translocation of Ca^{++} and other ions between intra- and extracellular pools is found in liver under O_2-limiting conditions (Farber et al., 1981), in acute renal failure in mammals (Trump et al., 1981), and can be presumed to occur generally in mammalian organs and tissues during O_2 lack. Although mechanisms remain to be clarified, such membrane failure might be facilitated by ATP-sensitive K^+ channels opening at the low intracellular [ATP] found during hypoxic stress (Spruce et al., 1985). In sharp contrast, in the brain and

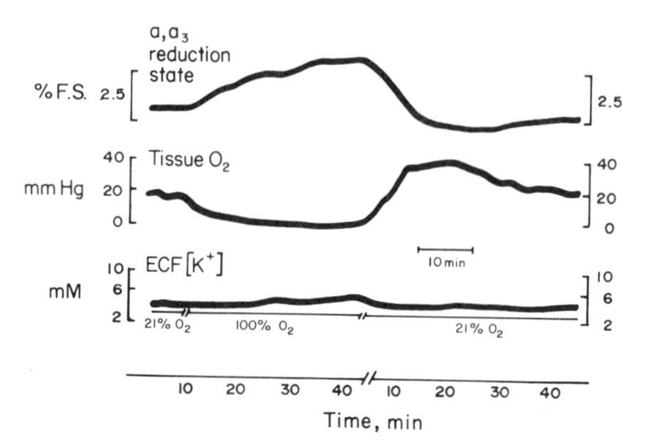

Figure 2.8 Changes in the redox state of cytochrome a,a_3, tissue oxygen tension, and extracellular concentration of potassium ions in turtle brain that resulted from switching the inspired gas mixture from 21% O_2 to 100% N_2. % F.S., percentage of the full-scale change in reduction state of cytochrome a,a_3 (in arbitrary units). From Hochachka (1986); copyright 1986 by the AAAS.

other organs of "good" ectothermic anaerobes during hypoxia, such failure of membrane function does not occur at all, or develops relatively slowly (Figure 2.8). As soon as continuous and high rates of ATP generation are reduced in mammalian tissues, intra- and extracellular ion concentration gradients are rapidly lost and tissue viability is at risk. In contrast, ionic concentration gradients do not fall to their thermodynamic equilibrium in tissues of ectothermic anaerobes despite metabolically depressed states in anoxia and proportionately *lower* ATP turnover rates (Sick et al., 1982; Surlykke, 1983). Obviously, something about cell membranes in facultatively anaerobic ectotherms is different; under hypoxic conditions, either these membranes are more impermeable to ions or ionic pumping capacity can pace thermodynamic drift to electrochemical equilibrium. Stabilized ion gradients almost certainly cannot be due to accelerated ion pumping because ATP turnover rates are lowered in the metabolically arrested states typical of animal anaerobes in anoxia. For this reason, we assume that the aforementioned membrane-based differences in the effects of hypoxia are mainly due to different permeability barriers and are an expression of a basic difference between cell membranes of hypoxia-tolerant eurythermic anaerobes and hypoxia-sensitive stenothermic ectotherms and endotherms.

Regulation of Membrane Permeability

Because moment-to-moment regulated changes in membrane permeabilities to ions are requisite for normal function in most tissues and organs, any large permeability differences between homologous cold-tolerant and cold-sensitive tissues must be due to fundamental specializations in the way membranes are structured or in their regulated function. In general, liquid–crystalline lipid bilayers are very permeable to water (which exhibits permeability coefficients of 10^{-2} to 10^{-4} cm sec^{-1}) but very impermeable to most small ions. Permeability coefficients of less than 10^{-10} cm sec^{-1} are commonly observed, but they can even be smaller; Na^+ and K^+ permeability coefficients, for example, are as small as 10^{-10} cm sec^{-1}. From what is currently known about membrane properties, basic differences in ion permeabilities could arise from (1) change in the phospholipid composition and consequently in functional properties of membranes, (2) change in the ratio of functional to nonfunctional ion-specific channels by change in the recruitment of channels from nonfunctional pools or "storage" sites, or (3) change in the abundance of "pores," that is, in the density of functional ion-specific protein channels.

The first of these, requiring adjustments at the level of bilayer composition, is probably of minimal importance, because a major function of bilayer adjustments is to lower the temperature for phase transition, thus allowing normal membrane fluidity and normal membrane functions despite lower cell temperatures. In fact, some adjustments in membrane-bound enzymes such as Na^+,K^+-ATPases may serve to facilitate function in the face of the new bilayer composition in the cold. In all animals such "homeoviscous adaptations" in effect contribute to functionally similar membrane functions at different cell temperatures and thus probably do not contribute heavily to any tissue permeability differences between eurythermic anaerobes and hypoxia-sensitive mammalian cells. Thus, we are left with the ion-channel options.

Ion Channel Densities and Membrane Permeability

In assessing this matter it is worth emphasizing, as Hille (1984) has done, that even if bilayer adjustments influence permeability properties, they presumably do so mainly by changing the properties of proteins called ion channels. It is widely held that ions move across

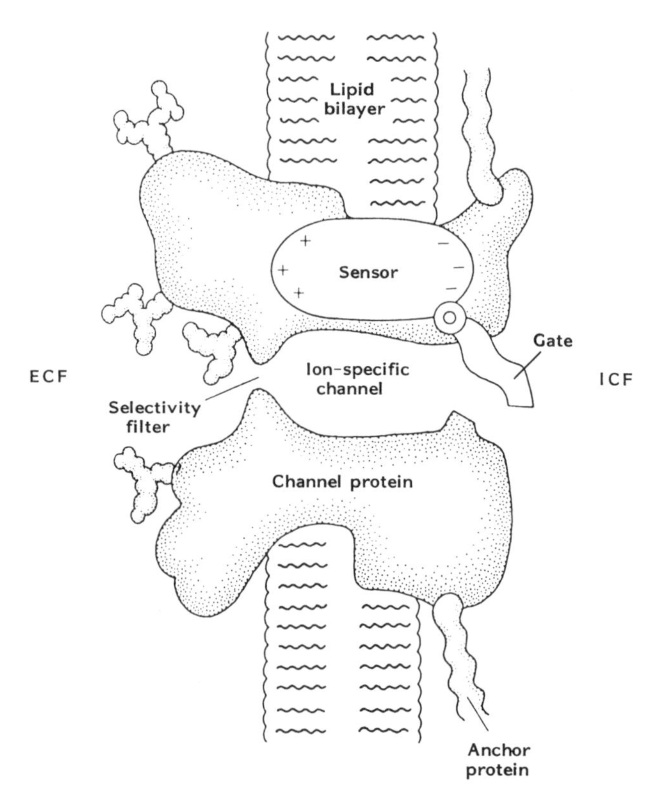

Figure 2.9 A working model of ion-specific channels. The channel is drawn as a transmembrane protein with a hole or pore through the center. The external surface of the molecule is glycosylated. The functional regions, selectivity filter, gate, and sensor are deduced from voltage-clamp experiments but are not yet understood at a molecular level. Modified from Hille (1984).

cell membranes through voltage-regulated or receptor-regulated aqueous pores or ion-specific channels, each having characteristic permeability, selectivity, and kinetics (Figure 2.9). In skeletal muscle and nerves, electrical excitation involves voltage- and time-dependent changes in Na^+, K^+, and Ca^{++} permeabilities through Na^+-, K^+-, and Ca^{++}-specific ion channels. At rest, Cl^- channels and other K^+ channels carry the dominant conductances. Perhaps the best understood of the identified ion channels, the Na^+ channel from at least two tissues is now described down to subunit composition and is being analyzed at the subunit level. When solubilized, the Na^+ channel protein from both muscle and nerve is about 316,000 daltons in size and consists of three nonidentical subunits. Transmembrane Na^+ flux is mediated by a hydrophilic pore containing a selective ion coordination site. By the selective use of neurotoxins, which bind with high affinity to specific sites on the Na^+ channel protein, it has become evident that Na^+ conductance through the channel is regulated or "gated" by controlling the rate and voltage-dependence of opening and closing of the channel (Catterall, 1984). In other voltage-dependent channels, opening and closing may also be linked to energy metabolism, for example, by ATP sensitivity (Spruce et al., 1985).

It is generally held that fundamental features such as protein size, oligomeric structure, subunit composition, ion coordination site, pore size, and neurotoxin binding sites are common to Na^+ channel proteins in all cell membranes, from different tissues and from different species; that is, like channel proteins in general, Na^+ channels in particular appear to be highly conservative. Channel density per μm^2 of membrane surface, however, varies between and within tissues, and in different functional states within the same tissue. The number of channels per square micrometer in rat brain, for example, ranges from about 100 in unmyelinated axons to possibly over 10,000 at the nodes of Ranvier in myelinated nerves, where the highest Na^+ fluxes are required (Catterall, 1984). In addition to the number of channels per unit area of membrane being regulated by long-term mechanisms, channel density may be modifiable on a moment-to-moment basis during transitions between different metabolic and physiological states. For example, recent studies of toad bladder indicate that antidiuretic hormone and aldosterone may influence the density of ion channels by controlled recruitment of preformed channels possibly stored in the cytosol (Palmer et al., 1982). Although additional studies on shorter-term regulation of numbers of functional channels in membranes clearly are needed, the evidence already available indicates that (at least in the long term) regulating the densities of functional, ion-specific channels may be a universal way of meeting tissue-, cell-, and ion-specific permeability requirements in different microenvironments or different metabolic states. In such event, the utilization of this strategy may explain the observed permeability differences in cell membranes between homologous tissues in "good" anaerobes versus hypoxia-sensitive endotherms. Furthermore, exactly the same ion channel density adjustments in principle are operable in membranes of mitochondria and of other intracellular organelles. For convenience, and because reduced membrane permeability in hypoxia-tolerant cells seems to derive from reduced number of functional ion channels, we term this the channel arrest concept of defense against hypoxia.

Predictions of the Channel Arrest Concept

The channel arrest concept explains several previously poorly understood observations rather well. In the first place, it explains (in fact, it predicts) the aforementioned comparative studies showing that, even if ion gradients do decline in hypoxia-tolerant ectotherms in hypoxia, the process is much slower than in endotherms, and that, in the extreme, ion gradients obviously remain stable after days of anoxia, even if ATP turnover rates are too low to explain the stability with ion pumping.

The channel arrest concept also helps to explain a rather long-standing paradox in environmental biochemistry and physiology of why some tissues and organs of animals at rest display O_2 consumption rates ($\dot{V}_{O_2}$) that vary with O_2 availability. Oxygen-conforming metabolic responses at the organ, tissue, and cell levels present us with an unresolved paradox that revolves around the problem of maintenance rate, or standard metabolic rate. In all O_2-conforming systems SMR obviously must decrease as O_2 availability decreases. This decrease in SMR may be achieved if the processes contributing to SMR also are turned down or off, or if their operational costs are paid for by anaerobic ATP-generating metabolism; in the latter event, a Pasteur effect would be necessarily observed. As already emphasized, however, O_2-conforming systems typically do not show a Pasteur effect. Indeed, where O_2-conforming patterns are most numerous (among invertebrate groups and ectothermic vertebrates), a *reversed* Pasteur effect is the rule. In the absence of a Pasteur effect, it is evident that the processes normally contributing to SMR in O_2-conforming organisms or cells are themselves necessarily downward adjusted as a function of O_2 availability. Because the most important single contribution to SMR probably is the cost of the ion pumping driven by ATP hydrolysis, maintenance of the same membrane potential state at reduced SMR and thus at reduced pumping rates necessarily requires reduced permeability (closing of ion-specific channels). Only when these adjustments are realized can O_2-conforming behavior be expressed at the cell and tissue level. We would like to emphasize that such behavior *and* such adjustments are most common and most striking in hypoxia-tolerant invertebrates and in hypoxia-tolerant endothermic tissues, as would be predicted by the channel arrest concept. When we turn the situation around, however (that is, move up the O_2 saturation curve, rather than down), we are confronted with the problem of why *increased* densities of functional ion channels develop as O_2 availability increases (as ATP synthesis rates increase). Even though we do not know the answer to this problem, an interesting possibility is that cells and membrane-bound organelles in effect become more reactive at higher channel densities. Ion fluxes through channels usually subserve specific biochemical or transduction functions, and these would presumably be favored at higher channel densities; hence high channel densities, high ATP turnover rates, and ATP-dependent ion channels (Spruce et al., 1985) may codevelop and be coadaptive. Conversely, we expect that a metabolically arrested organism, such as *Mytilus* or *Arenicola,* below critical P_{O_2} values, may be in relatively nonreactive states, analogous to anesthetized states. We do not know whether this adequately describes anoxic invertebrates, but the description empirically fits anoxic goldfish and turtles very well indeed. In this view, then, an additional cost of O_2 conformity or of metabolic arrest capacities may be reduced reactivities under O_2-limiting conditions. That this is an acceptable price for survival is indicated by the large number of animals sharing these strategies of adaptation to O_2 lack.

A third problem area that the channel arrest interpretation puts into somewhat clearer focus concerns the origins of endothermy. Recent comparisons of mammals and reptiles indicate (1) that ATP turnover rates and ouabain sensitivities of homologous tissues in mammals are about five times higher than in ectothermic reptiles, but (2) that the "leakiness" of cell membranes in mammals is also severalfold greater than in reptiles. Else (1984) and Hulbert and Else (1981) suggest that the latter explains the former; that is, one cost of endothermy is a higher rate of thermogenesis arising in part at least from "leaky" membranes and from the consequent necessity for higher ion pumping rates and higher ATP turnover rates. Leaky membranes thus may be adaptive in endothermic tissues because they are a part of an O_2-fueled biological furnace, whereas nonleaky membranes are adaptive in ectothermic anaerobes because they allow metabolic arrest without the risk of a breakdown in ion regulation and membrane functions generally. The implication that leaky membranes are usually necessary for endothermy but lead to an increased sensitivity to hypoxia in tissues of endotherms is exactly what is predicted by the channel arrest concept and appears to supply the element—stabilized membrane function—missing in earlier attempts to extend the hypoxia tolerance of hypoxia-sensitive tissues and cells.

Hypoxia-Tolerance: Metabolic Arrest Coupled with Channel Arrest

Now we have arrived at the kind of generalization we have been searching for—namely, that metabolic arrest must be coupled with a proportionate channel arrest to satisfy the minimal requirements for establishing in hypoxia-sensitive tissues the hypoxia tolerance of ectothermic anaerobes. Whereas this seems a simple enough strategy—one that is obviously utilized successfully by ectothermic anaerobes—is it experimentally realizable? Unfortunately, we do not know for sure; yet it is interesting that procedures directed precisely toward such a goal *have* been attempted. In these maneuvers (see Hearse et al., 1984; Fleckenstein et al., 1983, for example) the aim of intervention is to arrest the Ca^{++} channels of the plasma membrane and of intracellular organelles (SR and mitochondria, in particular) so as to block uncontrolled Ca^{++} fluxes. The intervention is designed, in effect, to block one or more of the steps leading to a rise in cytosolic [Ca^{++}] or to phospholipase activation as shown in Figure 2.6. Calcium channel arrest is thus being used as a first line of defense designed to protect hypoxia-sensitive tissues against O_2 lack, and it is often coupled with hypothermia-induced metabolic arrest. Although the application of this procedure to any mammalian tissue or organ so far investigated supplies a "protective" effect against hypoxia for short time periods, long-term protection appears still to be unachievable. The problem seems to be that hypothermia per se is damaging; and its disrupting effects in combination with hypoxia may well be exaggerated, because at the cell membrane the two stresses are disrupting for remarkably similar reasons (see Chapter 4 and Hochachka, 1986). So before further progress can be expected in protecting most mammalian tissues with these measures, metabolic arrest mechanisms other than hypothermia must be harnessable. To our knowledge, the only mammals who seem capable of utilizing the dual strategy of metabolic and channel arrest as protective measures against hypoxia are aquatic species during diving, which we shall analyze in the next chapter.

Summary

A simple and unifying picture of cellular mechanisms underlying tolerance to hypoxia emerges from the preceding analysis. In hypoxia-sensitive systems, an unavoidable depression in ATP synthesis rates due to curtailment of oxidative metabolism is potentially greater than any associated change in the passive leak of ions across membranes. In the absence of any additional adjustments, ATP synthesis rates cannot match ATP requirements for sustained stable membrane functions. This problem does not appear to be resolvable in anoxia-sensitive cells; thus during prolonged anoxic or ischemic exposure, ion gradients dissipate and intracellular Ca^{++} concentrations rise and activate membrane phospholipid hydrolysis in a process that ultimately leads to cell damage or cell death. Anoxia-tolerant cells avoid this Ca^{++}-mediated pathogenic process by maintaining low-permeability membranes (possibly by means of lower densities of ion-specific channels) so that the energy costs of ion pumping can be matched by the rates of ATP synthesis realizable under the metabolically depressed conditions of hypoxia. As O_2 availability declines, this channel-arrested (low permeability) state is coupled with a proportional metabolic arrest in "good" animal anaerobes, with metabolic arrest being achieved by at least three mechanisms. The first of these, termed O_2 conformity, is based on a true O_2 limitation to oxidative metabolism and correlates with partial depletion of mitochondrial adenylates. In the second mechanism, termed ETS arrest, mitochondrial ATP synthesis is fully blocked either because of total O_2 lack or because of essentially complete depletion of mitochondrial adenylates. The third mechanism takes over once the system is in complete anoxia; instead of making up the energetic shortfall due to O_2 lack (with a positive Pasteur effect), regulatory processes are harnessed to reverse the Pasteur effect, thus assuring low glycolytic flux despite anoxic conditions.

The degree of overall metabolic arrest varies in different kinds of hypoxia-adapted cells, tissues, and organisms, but is typically 5- to 20-fold. When these systems are under conditions of hypoxic stress, biological time therefore is extended (relative to clock time) by the same order of magnitude. The only endothermic examples we are aware of that (1) are able to sustain anywhere near this level of metabolic arrest and (2) are able to couple it with channel arrest as a dual defense strategy against potentially limiting O_2 availability are diving mammals and birds (Chapter 3).

3

Diving Mammals and Birds

At least since the time of Bert about a century ago, biologists have been intrigued by the effectiveness of O_2 management during diving in aquatic mammals and birds. Although all readers will be aware of the qualitative differences in breath-holding capacities between diving animals and terrestrial animals, most may not realize just how impressive this capacity can be. Some penguins, for example, are known to be able to dive in excess of 0.3 km for periods of up to about 20 min; Weddell seals in the Antarctic dive to depths of at least 0.6 km for periods as long as 1.2 hours, and deep-diving whales are capable of surpassing even these rather startling feats. As far as we know, in all aquatic mammals and birds the two basic metabolic requirements for any O_2-limited cells, tissues, or organs (conserving fermentable substrate and minimizing end product accumulations) during diving are met in part by a set of physiological reflexes that are termed the diving response and can be viewed as the animal's first line of defense against potential hypoxia. In laboratory studies of simulated diving, the diving response involves apnea, bradycardia, and peripheral vasoconstrictions, all of which are always observed. Metabolic effects of the diving response include (1) preferential redistribution of O_2 and blood-borne substrates to specific particularly needy organs and tissues, (2) increased accumulation of anaerobic end products such as lactate in hypoperfused regions of the body, concomitant with declining plasma glucose levels, and (3) a distinct lactate washout profile during recovery, with a concomitant hyperglycemia evident in the plasma (see Murphy et al., 1980).

Although some of these patterns seen during simulated diving were well understood by Scholander and Irving more than 40 years ago, controversy has now arisen concerning when, and even if, this reflex diving response is used during voluntary diving of aquatic mammals and birds. This controversy arose when biologists abstracted two characteristics of laboratory diving that are always elicited, then looked for them in voluntary diving animals operating in normal habitats (Kooyman et al., 1980; Kanwisher et al., 1981; Butler and Jones, 1982). The first of these (bradycardia), which is also most easily measured, was found to correlate rather unreliably with voluntary diving, often not being observed at all in short dives. The second characteristic chosen to assess the use of the diving response in nature (a lactate washout in postdiving recovery) is more difficult to measure in field settings and therefore is less frequently reported. Nevertheless, when it has been measured in Weddell seals, for example (Kooyman et al., 1980), it is rarely observed in short (feeding) dives (meaning, for this species, dives of less than about 20 min). These data were taken as evidence against use of the reflex response in voluntary diving animals and set the stage for a vigorous dialogue between field and laboratory biologists. We shall illustrate the issue by using data from studies of seals though the overall interpretation seems valid for all diving endotherms.

The Diving Response: Fact or Artifact

An important milestone in unraveling this tangle came from a recent interdisciplinary study of gray seals suggesting that a larger part of the aforementioned controversy arises from comparing a similar process (breath-hold diving) in the presence and absence of exercise (Castellini et al., 1985). Although both exercise and diving have been studied in aquatic mammals and birds, they have usually been examined separately; so it is not widely appreciated that the metabolic and physiological demands of these activities are in conflict. Nevertheless several fundamental and opposing requirements are raised by the two activities (Table 3.1), a finding that may readily explain why studies using only diving as a variable tend to favor the validity of the classic diving response, whereas studies involving both variables tend to favor an "exercise" model of diving, viewing diving as rather analogous to any other exercise, such as surface swimming, for example.

At least for the Weddell seal this controversy is now resolved, thanks to the development by R. D. Hill of a compact microcomputer backpack, which the study animal wears. The instrument allows monitoring of heart rate, swimming speed, diving depth, and diving duration and can operate a peristaltic withdrawal pump for obtaining sequential arterial blood samples (before and after injections of isotopically labeled metabolites) at any time during voluntary diving at sea.

In designing these studies (Guppy et al., 1986) the key insight was to recognize that the crux of the problem was sorting out the effects of two variables at once, since the general parameters used by all previous workers (swim speeds, heart rate, breathing rate,and the concentration profiles of plasma metabolites during and after diving) all are modifiable by diving, by exercise, or by both. These are thus insufficient to assess how one activity affects the other. The swim mill study of gray seals, on the other hand, developed realistic metabolic criteria to discriminate in the field between (1) a submerged animal exercising at varying intensity and (2) a diving animal utilizing classic or modified diving responses. These criteria depended on the observation that, following injection into the descending aorta (DA) of [^{14}C]glucose or [^{14}C]palmitate, the kinetics of replacement of labeled molecules with unlabeled molecules during rest and exercise are qualitatively and quantitatively different from the patterns observed during diving (Figure 3.1). When tested in field studies of voluntary diving Weddell seals, these criteria were adequate to settle the main issue at hand.

As shown in Figure 3.2 (for a 28-min dive with classic lactate washout profile; considered a long or "exploratory" dive), the specific activity (SA) for glucose does not decline along the smooth decay curve observed in the resting state (Figure 3.1). Instead, as the bolus

Table 3.1 Potentially conflicting demands of diving and sustained exercise[a]

Diving	Exercise
Low heart rate	Elevated heart rate
O_2 limitations	Aerobic conditions
Aerobic + anaerobic metabolic pathways	Mainly or solely aerobic pathways
Tissue-selective vasoconstrictions	Tissue-selective vasodilations
Whole-organism metabolic rate unchanged or minimally elevated	Metabolic rate elevated at maximum by an average of about 10-fold

a. See Guppy et al. (1986) and Castellini et al. (1985).

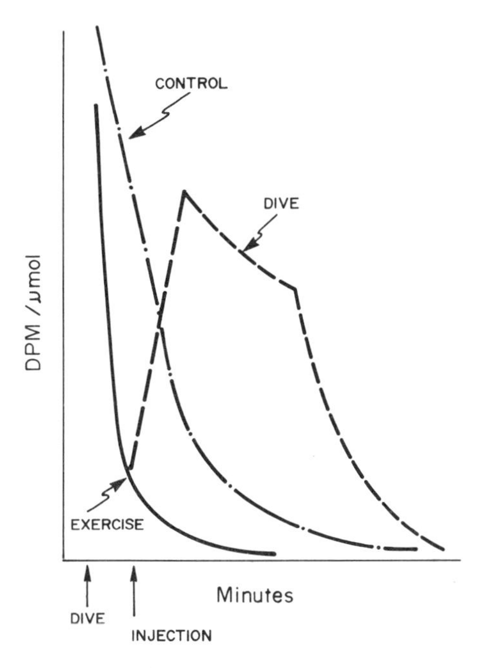

Figure 3.1 Diagrammatic representation of qualitatively different specific activity (SA) curves produced by two models of metabolic and physiological organization during voluntary diving of aquatic animals in their natural habitat. All exercise models of diving animals predict a curvilinear falloff in SA of metabolites injected after diving has started, a pattern analogous to that obtained during surface swimming exercise or during rest (or control state). In contrast, utilization of a true diving response (involving significant cardiovascular and flow adjustments) slows down the rate of "wash-in" and mixing of the labeled metabolites injected, thus generating more complex SA patterns with time. DPM, disintegrations per min. Based on data of Castellini et al. (1985).

is injected into the descending aorta toward vasoconstricted regions of the body, perfusion and mixing are so slow that a rising phase in SA of glucose is readily observed, and the SA does not reach a maximum value during the first half of the dive (Figure 3.2). After the dive, the decline of glucose SA is similar to or even faster than the control rate, a result indicating a high rate of glucose replacement (due both to exchange at tissues and to metabolism per se). This pattern is analogous to that observed during laboratory diving of the gray seal and similar to the patterns measured following injection of [^{3}H]palmitate or [^{14}C]lactate. As these kinds of long dives sustain a distinct bradycardia and a washout of lactate in postdiving recovery, the results are similar to laboratory dives. But most dives in nature, even in the Weddell seal, are short-term, "aerobic" dives—so termed because they show no lactate washout in recovery. Because the bradycardia observed in the aerobic dive is less than that observed in long dives, an extreme interpretation of these data would assume that the animal is in an exercise mode, not a diving mode, at such times. Thus the crucial tracer experiment required is for short dives, for all such "exercise" models of diving would require SA values for labeled metabolites to decrease in a simple hyperbolic pattern, as they do at rest or during exercise (Figure 3.1). Instead, the progressive changes in SA for glucose, palmitate, and lactate during short dives are qualitatively similar to those observed during longer exploratory diving (glucose data shown in Figure 3.2).

Quantitative differences are found between SA changes in short dives and those in long dives. In the first place, peak SA values are obtained earlier during short dives. Second, it is evident (from the low rate of change of SA in later portions of the dive when the label is fully equilibrated) that the replacement rates for all three metabolites

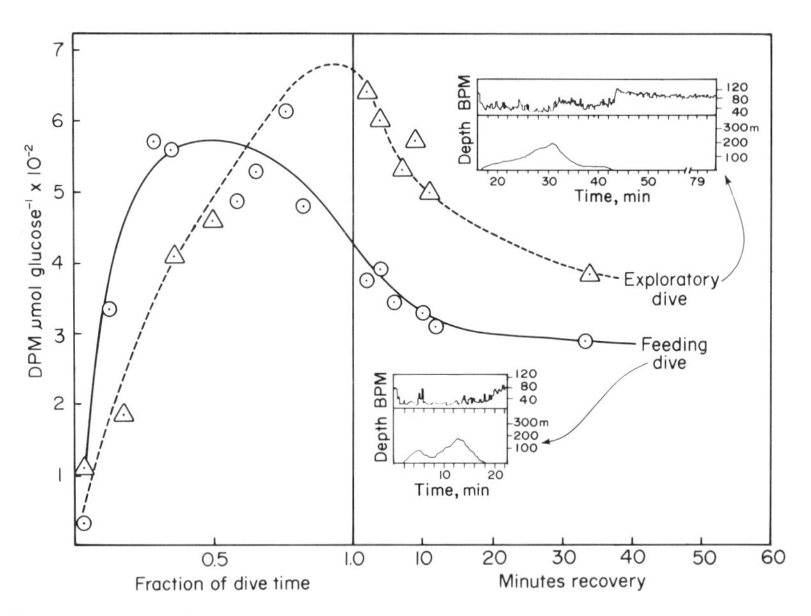

Figure 3.2 Complex pattern of change in SA of glucose during a 28-min exploratory dive compared to a shorter (17-min), feeding dive. Heart rates and depth of diving shown on the insets. From Guppy et al. (1986).

are greatly reduced (presumably because of reduced metabolic rates) during diving from those observed at rest or during recovery, despite the fact that the animal is exercising (swimming) at this time. Third, estimates of M_s values for these metabolites (the amount of each metabolite in the rapidly mixing pool) are similar to values obtained at rest, a result indicating that in both cases they are ultimately fully mixed and equilibrated in the seal's plasma volume and that the results cannot be faulted on the basis of some sort of regional compartmentalization of the injected labeled metabolites.

The Lactate Paradox

Even though the data imply many similarities between laboratory and voluntary diving, there are some important differences, particularly concerning lactate and swimming muscles. As first pointed out by Kooyman et al. (1980), in voluntary dives of short (less than 20 min) duration, a washout of lactate in recovery is not typically observed in Weddell seals (Figure 3.3), while in enforced dives as short as 5 minutes, this response leads to measurable lactate accumulation due to tissue hypoperfusion. The metabolic paradox, then, is why no lactate washout profile is seen following feeding dives at sea (Figure 3.3). One extreme possibility is that seals during short dives are merely in an exercising mode analogous to swimming at the surface, are nowhere O_2 limited, and are not dependent on the diving response to distribute a fixed amount of available O_2. But the data in the preceding section indicate that for all but the shortest dives this explanation is inadequate. It can also be ruled out simply because normoxic resting metabolic rates are high enough to consume all available O_2 supplies in about 20 min in a 450-kg seal. During voluntary diving, however, the seal is not resting, and its swimming muscles presumably require much of the "on board" O_2 supply, a situation that would favor development of O_2-limiting conditions in hypoperfused (nonworking) muscles and other peripheral tissues. Therefore we consider an alternative explanation of the paradox more likely. According to this view the plasma concentration of lactate in any given state is determined by the balance between entry and exit rates (fluxes termed R_1 and R_2 in Figure 3.4). Lactate accumulates if—and only if—R_2 is less than R_1, which is obviously the case in enforced diving. In voluntary diving this condition is not realized, presumably because the two diving modes differ most, not in lactate production rates, but

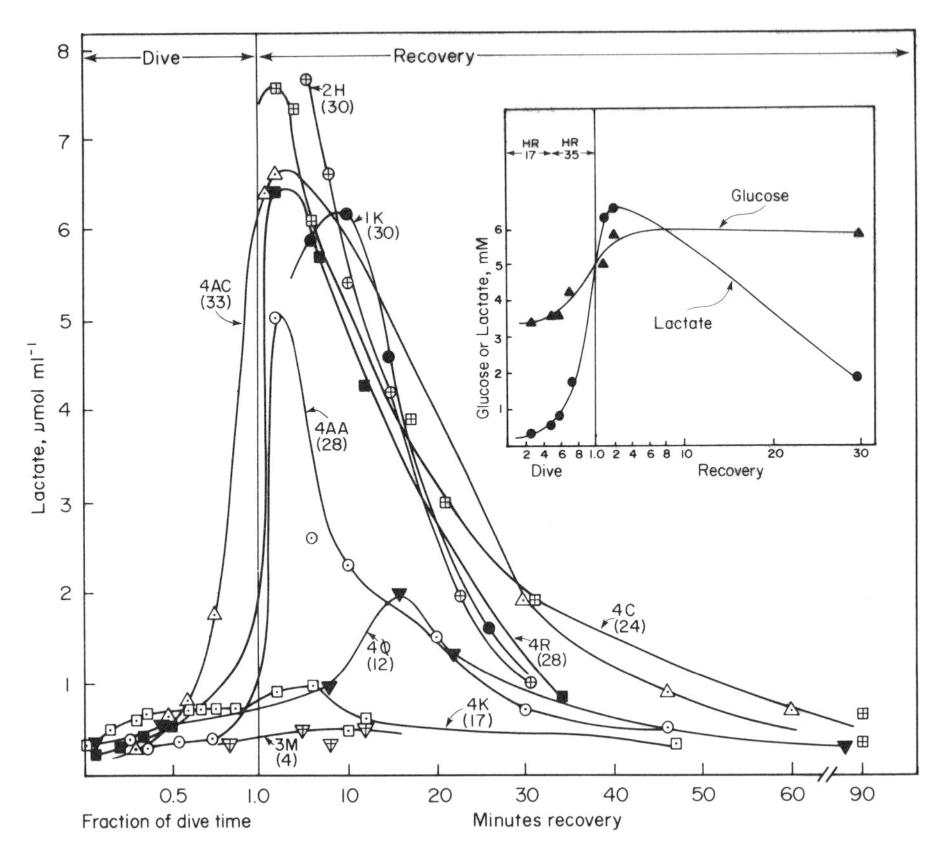

Figure 3.3 Plasma lactate concentrations during diving for various time periods and during recovery. Each dive is identified with a code, and its duration in minutes is given in parentheses. Following short dives a lactate washout usually is not observed, whereas following long dives the lactate washout measured is always substantially less than would be predicted if anaerobic glycolysis were to make up the aerobic energetic shortfall due to O_2 limitation. From Guppy et al. (1986).

in its disposition rates (R_2 in Figure 3.4). That is, seals during short voluntary dives must utilize lactate at rates, or at sites, not utilized as effectively during exploratory dives. One such site may be the liver, where the main fate of lactate in seals and other mammals is reconversion to glucose and glycogen, a function that may be more favorable during feeding than during long exploratory dives (when, as we shall see later, the liver may not be as well perfused). Another, qualitatively more important site of lactate metabolism, previously over-

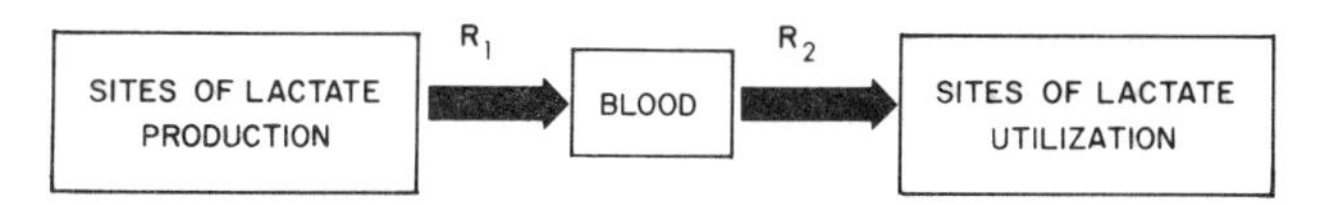

Figure 3.4 In any given metabolic state, the plasma concentration of a metabolite such as lactate is determined by the balance between entry rates into this compartment (R_1) and exit rates from it (R_2). During diving without exercise, $R_1 > R_2$ and consequently lactate accumulates. But during voluntary diving, which includes exercise, $R_1 = R_2$ or R_2 exceeds R_1, presumably because working skeletal muscle avidly utilizes lactate as a substrate during aerobic metabolism. Under these conditions no lactate would be expected to accumulate; none is observed.

looked in diving animals, is the swimming musculature. Whereas glycogen and fatty acids are most often the preferred fuels for aerobically working mammalian muscles, lactate is now known to be avidly utilized when made available. Furthermore, in vivo studies of integrated metabolism indicate that lactate utilization rates (especially reconversion rates to glucose in the liver and oxidation rates in working muscles) are strongly elevated during exercise in mammals (see Guppy et al.,1986, for literature in this area).

Is muscle energy metabolism during diving active enough to be consistent with such an interpretation? To assess this question in the Weddell seal, we can estimate muscle energy requirements during feeding dives from known "on board" supplies of O_2. For a 450-kg seal, with 135 kg of muscle and a partitioning of 75% of available (hemoglobin- and myoglobin-bound) O_2 for muscle work, a 30-min fully aerobic dive would be supported by a muscle metabolism consuming 1125 mmol O_2 (equivalent to turning over about 6750 mmol ATP). If all 135 kg of muscle were used for swimming, its metabolic rate would equal only 1.7 μmol ATP g^{-1} min^{-1}, which is about 1/12 the sustainable maximum for human muscle. This value is an upper-limit estimate, of course, for not all myoglobin-bound O_2 can be used for swimming, nor need the skeletal muscles receive fully 75% of the available O_2. However, this value is in the same range as that found from more direct measurements (Quist et al., 1986), so it seems fairly realistic. What is perhaps more relevant is that these rates of O_2-based muscle metabolism in the Weddell seal could utilize lactate at 20 mmol min^{-1}, a process that, coupled with some liver utilization, is easily high enough to explain why little lactate accumulates during feeding dives and why there is not enough left for a significant washout profile during recovery.

Low metabolic rates imply low flux rates of plasma metabolites such as palmitate, lactate, and glucose. For both voluntary and enforced dives, once the injected substrates are fully mixed into their respective unlabeled plasma pools, the rates of decline in SA values are proportional to the rates at which labeled substrate molecules are replaced by unlabeled ones, and these SA values clearly decline more slowly than during postdiving recovery or during rest (Figures 3.1 and 3.2). That is why we conclude that the replacement rates (R_a) for all three substrates so far tested decline during diving, even if the diving period is not extended long enough for calculating the areas under the SA curves and, thus, accurate R values (Guppy et al., 1986).

Because in other animals the R_a is determined by perfusion and plasma substrate availability, reduced metabolite fluxes during diving would also be expected simply because of low perfusion rates and, in the case of glucose, reduced availability. Perfusion, however, is linked to heart rate and cardiac output, and the latter appears to vary according to swimming speed; that is, increased cardiac output is accounted for by increased perfusion to working muscles, according to the energy needs of swimming. Thus even if the replacement rates of plasma substrates are reduced for the whole organism, the fluxes of substrates and O_2 to working muscles could well be elevated in proportion to an exercise-induced increase of perfusion, as in terrestrial animals.

Resolving Conflicting Demands of Diving and Exercise

How might the balance between diving and exercise demands be achieved? In physiological terms, this problem can be rephrased in terms of how the conflicting needs of diving-dependent vasoconstriction and of exercise-mediated vasodilation can be resolved at the levels of working muscle. Although we do not know how this conflict is resolved, two clues are available from heart rate data. First, diving bradycardia may be graded according to diving mode; so if vasoconstrictions are graded in a similar manner, the circulation to muscles may be sufficient to account for an adequate flux of O_2 and substrates. A second clue comes from recent data showing periodic relaxations of bradycardia during both feeding and exploratory dives (Guppy et al., 1986). If these are coupled to simultaneous relaxation of vasoconstriction to swimming muscles, a mechanism would be made available to periodically perfuse vasoconstricted muscles and so supply a pathway for O_2 and substrate fluxes from sites of entry into plasma to sites of utilization. However, additional measurements of muscle flow and metabolism are obviously needed before such alternatives can be experimentally assessed.

For seals at rest, then, the specific activities of labeled palmitate, lactate, and glucose injected into the DA decline along fairly smooth, nearly hyperbolic curves. Such decay patterns of SA over time are best described as multiexponential curves, because in these kinds of experiments the injected metabolites mix rapidly and completely in the plasma but mix much more slowly in other body compartments. However, the patterns obtained are theoretically expected and widely observed in injection experiments in all animals so far studied, including gray seals at rest and during exercise. An essential point is that *all* versions of voluntary diving that assume that the animal is essentially in an exercise mode require similar SA patterns during voluntary diving, whereas models assuming the animal to be in a true diving mode would predict drastically modified patterns, similar to those noted during laboratory diving in the gray seal (Figure 3.1). Direct measurements in voluntarily diving Weddell seals made on samples remotely drawn with a microcomputer-controlled peristaltic pump are all inconsistent with the predictions of an "exercise" model of diving. On the other hand, these are entirely consistent with the alternative model of diving; in this view, the simplest interpretation of the data is that mixing and equilibration of labeled and unlabeled metabolites are drastically slowed down because the labeled substrates are injected into the descending arota and thus toward the vasoconstricted regions of the body. For this reason, in short dives at least several minutes are required for injected metabolites to be fully equilibrated in the available plasma pools, whereas in long, exploratory diving even more equilibration time is required. In both diving modes, however, the diving response is clearly utilized as a means of conserving O_2 and substrates for the animal's most needy tissues (heart, lung, brain, and locomotory muscles).

The Spleen as a SCUBA Tank

Studies of blood gases, hematocrit, and hemoglobin lead to an important extension of the diving response beyond the direct cardiovascular adjustments to the control of O_2 content and O_2 delivery during diving. During the first 10–15 min of both long and short voluntary dives, blood hemoglobin (Hb) content increases by nearly 60% (hematocrit increases from about 40% to as high as 65%). In terrestrial species, such as the sheep and horse, the spleen is known to function as a dynamic red blood cell (RBC) reservoir, at rest containing 26 and 52% of the total RBC mass, respectively. Excitement, exercise, or catecholamines cause the spleen to contract and increase the hematocrit by 25% or even more. A similar mechanism appears to operate during diving, when the Weddell seal takes advantage of sympathetic vasoconstriction of its peripheral vasculature to induce constriction of its very large spleen (Qvist et al., 1986). In this way, over the first 10–15 min of diving, the seal gradually injects more and more of the approximately 50% fraction of RBC mass that is initially stored (oxygenated) in the spleen. In effect the spleen behaves like a regulated SCUBA tank in a process so finely tuned (O_2 uptake balancing O_2 delivery) that for the first 10–15 min of diving the O_2 content of the blood, which would otherwise drop as a result of tissue O_2 consumption, is maintained at a constant level by precise pacing of injection rates of oxygenated RBC. Once the full RBC mass is circulating, O_2 content declines as anticipated; because the lungs are collapsed and nonfunctional in gas exchange, the decline can be used to calculate metabolic rate.

Metabolic Rate during Diving

To illustrate metabolic conditions during the dive state, we shall quantitatively examine a hypothetical 30-min dive in an adult Weddell seal. During the first 15 min, the vertical velocity averages 0.23 m sec^{-1}, the heart rate averages 15 beats min^{-1} (BPM), and an initial blood O_2 content of 345 ml O_2 L^{-1} is reduced to 230 ml L^{-1} (conditions frequently observed). Assuming a blood volume of 60 L in a 450-kg seal, the O_2 uptake from the blood is estimated to be 6900 ml, of which about 1815 ml O_2 is consumed by heart, lung, and brain metabolism. If all the rest of the available O_2 is utilized by working muscle, its maximum aerobic metabolic rate would be 5085 ml O_2 over the first 15 min of diving. If all 135 kg of skeletal muscle are utilized in swimming, the mass specific rate of ATP turnover would be 0.75 mmol ATP kg^{-1} min^{-1}; under these conditions, whole-organism O_2 uptake rate ($\dot{V}_{O_2}$) equals 460 ml O_2 min^{-1}, or about 24% of a resting $\dot{V}_{O_2}$ of 1900 ml O_2 min^{-1}. In the second half of such a dive, heart rate typically increases, as does swim speed. At a heart rate of 25 BPM, the heart, lung, and brain utilize about 3000 ml O_2 in 15 min. If the remaining blood O_2 supplies, plus all myoglobin-bound O_2 stores (which in 450-kg Weddell seals are estimated at about 10,000 ml O_2) were used to support the work of 135 kg of muscle, the muscle metabolic rate would be 4.9 ml O_2 kg^{-1} min^{-1}, equivalent to an ATP turnover rate of about 1.5 mmol ATP kg $muscle^{-1}$ min^{-1}, at a $\dot{V}_{O_2}$ for a 450-kg seal of 863 ml O_2 min^{-1}, or about 45% of SMR. These metabolic rates could vary and admittedly are only approximations (as before, these are probably too high because not all myoglobin-bound O_2 could be depleted under these conditions, and the assumed O_2 content at the end of diving is somewhat lower than average). Yet two instructive insights arise from the calculations. First, heart rate and swimming velocity during diving should be, and *are*, related, so it is a fair assumption that working skeletal muscles are powered by oxidative metabolism. Second and more important from our perspective, the energy requirements of the diving Weddell seal obviously are low—surprisingly low—both in terms of mammalian working muscles and in terms of the whole-organism metabolism in other states (Table 3.2).

Metabolic Arrest in Hypoperfused Tissues and Organs

How does a 450-kg Weddell seal operate on only fractions (24 and 45%, for example) of resting metabolic rate during descent and as-

Table 3.2 Estimated metabolic rates for adult Weddell seals under different metabolic states

State	$\dot{V}_{O_2}$ (ml O_2 450 $kg^{-1}min^{-1}$)	Calculated[a] mmol lactate formed per 30 min by equivalent anaerobic metabolism	Effective[b] lactate concentration (mmol kg^{-1})
Resting	1,900	17,100	38
Diving			
25 BPM	863	7,767	17
15 BPM	460	4,140	9
Surface swimming[c]	3,800	34,200	76
Maximum exercise[d]	11,400	102,600	228

a. Assuming that the energetic shortfall on complete cessation of a given oxidative rate is made up fully by anaerobic fermentation of glucose to lactate.

b. Assuming lactate is equally distributed throughout a 450-kg body.

c. Assuming 2-fold activated metabolism as in gray seals at 2 m sec^{-1} observed by Castellini et al. (1985).

d. Assuming a maximum aerobic scope for activity of 6-fold, which is typical for seals (see Castellini et al., 1985; Elsner and Gooden, 1983).

cent, respectively? One possibility is that it makes up the energetic shortfall with anaerobic glycolysis in hypoperfused organs and tissues (that is, it activates the classic Pasteur effect in these tissues). Because resting $\dot{V}_{O_2}$ is known, it can be shown that making up the energetic shortfall by anaerobic glycolysis in a 450-kg seal would lead in 30 min to an unusual accumulation of lactate, in excess of about 20 mmol kg^{-1} when completely equilibrated throughout the 450-kg body. Because from gray seal studies we know that surface swimming at comparable speeds leads to a 2-fold increase in $\dot{V}_{O_2}$, the actual energetic shortfall might be up to 2-fold higher, in which event making it up by anaerobic glycolysis would lead to about 40 mmol kg^{-1} throughout the animal's body. It is important to emphasize that this amount of lactate is much higher than ever observed in voluntary dives of such duration at sea. As in ectothermic anaerobes, the mystery of the missing lactate appears to be resolved by a reversed Pasteur effect, which is presumably sustained somewhere in the hypoperfused regions of the animal's body whenever the diving reflex is employed, in effect allowing energy metabolism to be switched down well below resting rates.

From studies of redistribution of cardiac output during diving, using either flow meters (to directly and continuously monitor flow to a given region) or microspheres (to monitor average flow under specified conditions), it is evident that most parts of the animal's body sustain significant hypoperfusion (Elsner and Gooden, 1983; Zapol et al., 1979). Except for the central organs and, in voluntary diving, the skeletal muscles used for swimming, all organs and tissues thus far tested sustain a significant reduction in perfusion and thus in delivery of O_2 and substrates. As emphasized in Chapter 1, metabolic rate commonly falls if O_2 delivery is too low. Thus, in a general sense, it is readily understandable why the overall metabolic rate of the organism may be reduced under these conditions. For two organs (liver and kidney), the background information seems solid enough to make some rather insightful calculations of the degree of metabolic arrest that is possible during diving.

Kidney Hypoperfusion and Metabolic Arrest Capacities

For over two decades now it has been well known that during enforced diving renal blood flow and renal functions can be significantly reduced. At the limit, the maximum response apparently leads to a *complete* cessation of renal blood flow and thus *complete arrest of oxidative metabolism* (see Murdaugh et al., 1961, for an example of this effect). In short, enforced dives of the Weddell seal, the drop in perfusion is only about 10-fold, from about 3 to 0.3 ml blood g^{-1} min^{-1} (Zapol et al., 1979). Similar changes are now believed to occur in voluntary diving. To assess kidney function during voluntary diving at sea, tracer quantities of [^{3}H]paraaminohippuric acid (PAH) and [^{14}C]inulin can be injected via a DA catheter while the seals are at rest or at the beginning of diving. After mixing, PAH clearance from the plasma is considered kidney-specific, with clearance rates being proportional to renal perfusion rates. In resting Weddell seals, plasma PAH clearance curves are hyperbolic with a $T_{1/2}$ (time to 50% clearance) equal to 8 min. The effects of exploratory diving on the wash-in of [^{3}H]PAH injected in the aorta are dramatic: instead of a rapid clearance of labeled PAH, circulatory mixing is incomplete even after 10 min into the dive, a finding strongly indicative of diving vasoconstriction. In long voluntary dives it appears that little or no PAH clearance occurs until the diving response is relaxed at the end of the diving period, at which time clearance patterns are similar to those occurring at rest. Similar results are obtained if [^{14}C]inulin is injected to measure the glomerular filtration rate (GFR). During long-term or exploratory diving, equilibrium is again delayed, a finding indicating vasoconstriction with minimal inulin clearance from aortic plasma evident until recovery. At this point the GFR rapidly returns to normal, and the [^{14}C]inulin clearance curve is almost identical to those in Weddell seals at rest. Thus we conclude that most, and perhaps all, renal functions are curtailed in voluntary diving mainly because of hypoperfusion (Guppy et al., 1986).

To understand how these perfusion changes are related to metabolism, recall that renal oxidative metabolic rate is the sum of a small but constant basal O_2 uptake rate (SMR) and a variable suprabasal or functional O_2 uptake rate. Kidney SMRs are defined as the O_2 uptake rates of nonfiltering and thus nonreabsorbing kidneys; in several terrestrial species (rats, cats, dogs, man) they are found to be remarkably similar, about 1 μmol O_2 g^{-1} min^{-1}. In the same spectrum of terrestrial species, the functional metabolic rates of the kidneys average about 30 times higher (Table 3.3) and are proportional to renal blood flow. If we assume that this proportionality is retained in seal kidneys during diving, then their O_2-fueled metabolic rate drops by anywhere from 1⁄10 of normal to zero, a result implying a reversed Pasteur effect, minimally of 2- to 4-fold and maximally of orders of magnitude (Table 3.3).

Because this degree of metabolic arrest is equivalent to a proportionate slowing of biological time, we would predict that the hypoxia tolerance of kidneys in diving animals would be similarly expanded. The most convincing data confirming this prediction are those of Elsner and his coworkers on isolated, perfused seal kidneys (Figure 3.5). Following extreme insults (initial ischemia during tissue collection, a subsequent 50-min period of hypothermia, a further 60-min stabilization period of perfusion with physiological saline, then finally a 60-min anoxic period induced by curtailing perfusion), the oxidative metabolic rate of the kidney returns to normal almost instantly, as do most of the measured renal functions. A similar period of ischemia brings about irreversible damage, both to metabolism and to renal function in kidneys from terrestrial species (Hochachka, 1986), a result attesting to the improved hypoxia tolerance of kidneys in divers such as seals.

Table 3.3 Calculated metabolic rates of kidneys of Weddell seals under different activity states

State	Oxidative metabolic rate (μmol ATP g^{-1} min^{-1})	Lactate formed per 30 min by equivalent anaerobic metabolism (μmol g^{-1})
Basal[a]	6	180
Reabsorbing[b]	15–30	450–900
Vasoconstricted in diving		
Short-term[c]	1.5–3	45–90
Maximum response[d]	0	0

a. Assuming basal O_2 uptake of nonfiltering, nonreabsorbing seal kidney is similar to that reported for terrestrial species (see Lassen et al., 1961; Gotshall et al., 1983).

b. Lower value assumes metabolic rate reported by Halasz et al. (1974) for isolated, perfused harbor seal kidneys measured following an initial ischemia during collection of variable duration, a 50-min cold ischemia, then a 60–90 min stabilization–perfusion period. This value is one of the lowest observed, about half the higher values typically reported for kidneys from terrestrial species (Lassen et al., 1961; Gotshall et al., 1983).

c. Assuming that the O_2 uptake of the kidney varies with perfusion, as in terrestrial species, and that during diving the flow rate is reduced from 3 ml g^{-1} min^{-1} to 0.3 ml g^{-1} min^{-1}, as observed by Zapol et al. (1979) for this species.

d. Assuming that renal blood flow is completely curtailed, as in a maximum diving response of the harbor seal (Murdaugh et al., 1961).

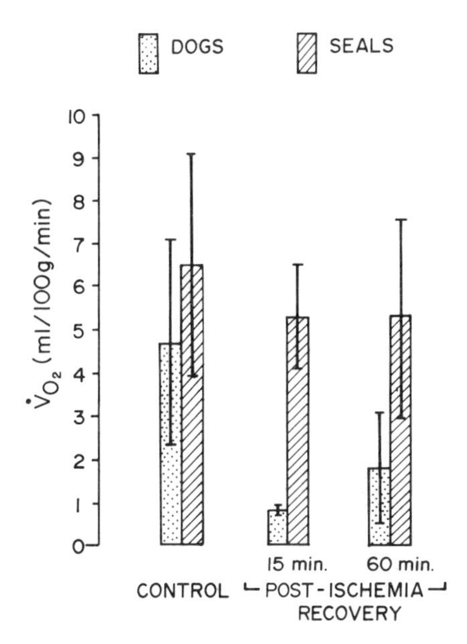

Figure 3.5 O_2 uptake by isolated, perfused dog and seal kidneys in control state and in a postischemic state following a 60-min anoxic (no flow) period. These postischemia recovery patterns parallel changes in renal blood flow (given in brackets in ml g^{-1} min^{-1}). Modified from Halasz et al. (1974).

Liver Hypoperfusion and Metabolic Arrest during Diving

As in kidney studies, two separate approaches (using flow meters and microspheres, the former by Elsner and his coworkers and the latter by Zapol and his group) clearly establish that blood flow to the liver during enforced diving is strongly curtailed. In the Weddell seal under laboratory diving conditions, hepatic arterial flow (measured by microspheres introduced into the left ventricle) is decreased to less than 1/20 of resting values. Similar degrees of liver hypoperfusion are obtained from flow meter measurements and also appear rather likely in voluntary diving.

To assess liver function during voluntary diving, cholic acid can be used as an organ-specific test compound. In such tests on resting seals (Figure 3.6), cholic acid is quickly cleared from plasma ($T_{1/2}$ = 5.6 min). The impact of a 12-min dive (without a lactate washout) on cholic acid clearance is dramatic (Figure 3.6, inset). As for all other aortic injections during diving, the equilibration and mixing of

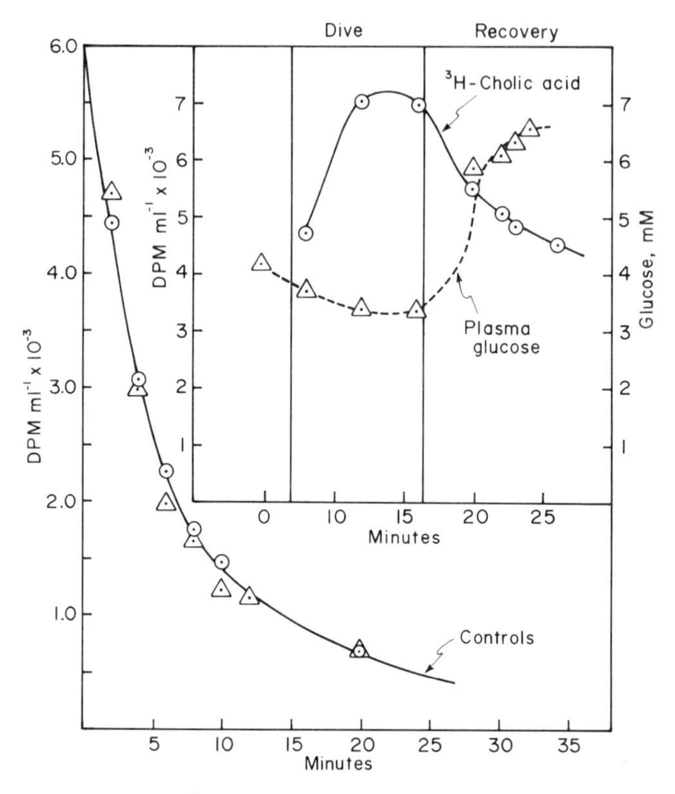

Figure 3.6 Clearance of [^{3}H]cholic acid from plasma of two Weddell seals at rest with values standardized to the same (27 L) plasma volume. Volumes of distribution for cholate for the two seals were 29 and 45 L compared to plasma volumes of 27 and 51 L, respectively. *Inset:* Effect of diving on cholic acid clearance in the Weddell seal. In this experiment, the bolus was injected at the beginning of a dive that lasted 3.5 min; the animal resurfaced for about 2.5 min, then entered another short (exploratory) dive lasting 12 min, during and following which sequential blood samples were obtained. From Guppy et al. (1986).

labeled cholic acid is delayed, so peak activity values (DPM ml^{-1}) are not reached until 5–7 min after initiation of diving. Although qualitatively similar to the kidney function tests, liver function tests differ in one important way: with cholic acid, equilibration and complete mixing appear to be attained well before the end of the short, feeding dive; hence the peak values of DPM ml^{-1} are obtained *during* the dive (Figure 3.5, inset). For at least 6 min after this peak is obtained, there is no evidence of [^{3}H]cholate clearance, although at rest this is sufficient time to halve the peak DPM ml^{-1} (Figure 3.6). These results imply better mixing within the plasma volume and less intense vasoconstriction of the liver during short dives; nonetheless they suggest that liver function is curtailed during short dives. As expected, the cholate clearance rates increase sharply when the dive ends. Simultaneously, plasma glucose levels rise, another indication that liver function is returned rapidly to normal. From the preceding results, we conclude that liver physiological functions are drastically reduced during simulated and voluntary diving.

The question now arising concerns metabolic consequences. As in the case of the kidney, the oxidative metabolic rate of the liver in all terrestrial mammals so far studied is very similar (Table 3.4), and these rates also drop off as perfusion decreases (in liver ischemia, for example). Assuming the same relationship is maintained for the liver during diving in seals, it is evident that the metabolic rate would be 0.1–0.05 of normal. This turns out to be a true metabolic arrest. If the energetic shortfall were to be made up by anaerobic glycolysis, the amount of lactate that would accumulate in the liver at the end of 30 min of diving would be prohibitively high (Table 3.4), whereas the lower estimate of ATP turnover rate could easily be accounted for by glycolysis. This calculation also indicates that the reversed Pasteur effect is of large magnitude: if it were not utilized, up to 20 times more lactate would be found than is otherwise required.

Metabolic arrests in diving animals profound enough to reduce energy metabolism to levels ranging from 10 to 60% of normoxic resting rates have been noted before in several species, usually under laboratory conditions (Elsner and Gooden, 1983). Thus the estimates

Table 3.4 Calculated metabolic rate of liver during rest and during diving in the Weddell seal

State	Metabolic rate (μmol ATP g^{-1} min^{-1})	Lactate formed per 30-min dive with equivalent anaerobic metabolism (μmol g^{-1})
Rest[a]	15–20	450–600
Diving[b]	1	30

a. Assuming that the liver metabolic rate in the Weddell seal is equivalent to that in terrestrial mammals (rat, cat, dog, lamb, man); see, for example, Edelstone et al. (1984), Lutz et al. (1975), and Lautt (1976). Also see Figure 2.1.

b. Assuming the liver metabolic rate declines according to the decline in perfusion (Zapol et al., 1979; Elsner and Gooden, 1983; Butler and Jones, 1982).

for the Weddell seal are unique only in that they also apply to a species diving voluntarily at sea. For this reason it appears that two conclusions can be made rather confidently. First, marine mammals use the diving response to precisely regulate the rates and sites of utilization of fixed amounts of O_2 over fixed periods of diving time. Second, metabolic arrest mechanisms (apparently involving a reversed Pasteur effect) are activated in hypoperfused tissues and organs to minimize overdepletion of carbohydrate in inefficient fermentations and to minimize end product accumulation. Of the two best defense strategies used by ectothermic anaerobes, at least one (metabolic arrest) can, therefore, also be harnessed by endothermic marine mammals.

Metabolically Arrested Tissues during Diving Are Also Probably Channel Arrested

What about the second strategy, the coupling of low-permeability membrane functions to metabolic arrest capacities? In considering this problem, we are faced with a conundrum similar to that encountered in our comparison of hypoxia-sensitive and hypoxia-tolerant systems: What happens to the dominant processes contributing to maintenance metabolism or SMR when SMR is turned down? This question can be rephrased in terms of how ion leakage rates down diffusion gradients are held in balance by ion pumping rates when the former are presumably uninfluenced by hypoperfusion while the latter are potentially strongly inhibited as O_2-fueled ATP synthesis rates decline toward zero. In principle, this problem can be solved either by compensation by anaerobic glycolysis for the slack-off in O_2-based ATP synthesis or by maintenance of low permeability cell membranes (low densities of functional ion-specific channels). All of the preceding data on kidney and liver (particularly the widely accepted need for metabolic arrest) argue cogently against compensation by anaerobic glycolysis; thus we are left with the alternative of low permeability. At least tentatively, we conclude that organs such as liver and kidney (and presumably other hypoperfused tissues of the seal's body during diving as well) either compensate for reduced ATP-dependent ion-pumping capacities during the hypometabolic hypoperfused diving state by reducing the densities of functional channels per square unit of cell membrane surface in proportion to the declining metabolic rate, or they maintain the required low channel densities all the time and are thus not stressed by the drop in ion pumping capacities. The net effect of either mechanism would be the same, that is, maintenance of the ratio of leak rates to pumping rates at unity even during metabolic arrest.

This mechanism could account for the finding that hypoperfused organs such as the liver and kidney can sustain hundreds of bouts per day of partial to complete metabolic arrest without any measurable decline in their physiological performance capacities. The same mechanism may also enable the seal kidney, when made fully anoxic for periods as long as an hour, to fully and instantaneously recover, not only metabolic rates, but also essentially all normal renal functions despite an insult that leaves the homologous organ of terrestrial species irreversibly damaged as a result of decoupling of metabolic and membrane processes. And finally, switching-down some tissues is a mechanism that could account for the observation that field metabolic rates of Weddell seals (from data that represent averaging over many hours per day of diving to great depth with interdive recovery periods) do not exceed the resting metabolic rates of seals, a paradox that has been perplexing to workers in the field and has not been previously explained adequately (Guppy et al., 1986; Kooyman, 1981).

In at least one group of tissues and organs and in one group of endotherms, then, the defense strategies against hypoxia that are widely used in ectothermic anaerobes (coupling metabolic and channel arrest) appear harnessable. When the underlying mechanisms are better understood, they may serve as a point of departure for further refining a potentially exciting intervention strategy for more applied problems.

Summary

During diving in aquatic mammals and birds two basic metabolic requirements for any tissues or organs that need to sustain periodic O_2 lack (conserving fermentable substrate and minimizing end product accumulation) are met in part by a set of physiological reflexes termed the diving response and are in effect the first line of defense against hypoxia. Under laboratory or simulated diving conditions the physiological reflexes involved are apnea, bradycardia, and peripheral vasoconstriction, all of which are observed in all diving animals. The metabolic effects of these reflexes include (1) preferential redistribution of blood-borne substrates and O_2, (2) reduced rates of substrate flux through the plasma compartment and a generalized reduction in metabolic rates of hypoperfused tissues, and (3) accumulation of anaerobic end products like lactate, with a distinct lactate washout evident during postdiving recovery. At least in the Weddell seal this strategy is now known to be utilized in voluntary diving at sea, but the response is necessarily modified to accommodate potentially conflicting demands of diving and swimming exercise. The main modification appears to involve skeletal muscles used in swimming, which, because of their high energy requirements, must be powered by a largely aerobic mechanism. Thus they must necessarily remain perfused at rates proportional to swimming velocity (which is why heart rates are adjusted to swimming velocity, the increased cardiac output presumably being mainly directed toward the swimming musculature). The required regulation of O_2 delivery is achieved in part by a well-paced release of oxygenated RBC, stored at the beginning of the dive in the spleen. The main metabolic difference between labortory and voluntary diving is that in voluntary diving working muscles serve as an effective sink for lactate; thus the entry rates of lactate into the plasma can be balanced by exit rates from the plasma. The maintenance of this balance means that no excess lactate remains for a lactate washout in postdiving exercise except after long, exploratory diving. Even in the long dives, however, the amount of lactate formed is far less than would be expected if the energetic shortfall due to hypoperfusion and O_2 lack were made up by anaerobic glycolysis. The missing lactate, here as in the ectothermic anaerobes discussed in Chapter 2, is accounted for by the utilization of a reversed Pasteur effect. Consequently, during diving most, if not all, hypoperfused tissues sustain a metabolic arrest of variable degree. In the kidney and liver the degree of metabolic arrest can be estimated because in these organs the rates of O_2-fueled ATP turnover are proportional to perfusion rates. Under laboratory diving conditions, perfusion rates of these organs are only 0.1–0.05 of normoxic values, and at the limit can be fully curtailed; O_2-based metabolic rates drop accordingly. Because of the high ATP turnover rates of normoxic liver and kidneys, the energetic shortfall again cannot be made up by glycolysis, so the true metabolic arrest is close to that estimated from the decline in oxidative metabolism. Biological time is accordingly slowed down or extended by factors of 10 to 20, relative to clock time. Under these conditions, a 30-min dive is equivalent to a 1.5- to 3-min vasoconstriction of these organs at normoxic metabolic rates.

4

Ectothermic Hibernators

In all vertebrate ectotherms (fishes, amphibians, and reptiles) body temperature (T_b) is dependent to some extent upon the ambient temperature. This dependence varies with the physical nature of the environment (it is much more difficult to conserve heat in water than it is in air) and naturally is more biologically significant in environments where temperature extremes or large temperature fluctuations occur—environments in which vertebrate ectotherms are found. Reptiles and amphibians are found from 71°N to Tierra del Fuego in the south, and fishes range virtually from pole to pole (Darlington, 1957). More specifically, the adder *Vipera berus* is found in Europe at 66°N (Cloudsley-Thompson, 1971), snapping turtles hibernate at 5°C (Gatten, 1978), the common frog winters in northern Finland (65°N) (Pasanen and Koskela, 1974), and American eels inhabit waters that range in temperature from 5° to 20°C (Walsh et al., 1983). Consequently, many of these vertebrates each year encounter a situation in which their T_b is so low as to prohibit such functions as movement, reproduction, feeding, and digestion. During this period these animals secrete themselves away in a variety of places, cease all obvious activity, and remain this way until temperatures begin to rise again in the spring—a period that can last 7–8 months! This period—cold torpor—is a fascinating one and raises some instructive questions. For instance, do these organisms maximize the effect of the temperature decrease; and if so, how? What are the metabolic problems that *must* arise during a 20°–30°C change in T_b and what are the solutions that have just as surely evolved to neutralize these problems? In the remainder of this chapter we will attempt to answer these questions and to show that torpor in ectothermic vertebrates is an active, organized, and concerted process during which time is put on 'hold' while heat, and in some cases O_2, is at limiting levels.

Inverse Temperature Compensation

The effect of temperature on any process can be described using the term Q_{10}, which is defined as the change in the rate of the process over a 10°C change in temperature. The Q_{10} for most chemical reactions lies between 2.0 and 3.0. The Q_{10} of chemical reactions that occur in organisms, as part of the metabolic process, can be altered, by various means, as part of the mechanism by which ectotherms adapt to temperature changes. These alterations can vary with season and with the range over which the temperature is changing. Thus some ectotherms, by perturbing the Q_{10} of reactions, may have a metabolic rate that is invariant with temperature over a 20°C range. Others may not compensate at all, while many vertebrate ectotherms that become torpid demonstrate inverse temperature compensation (Hazel and Prosser, 1974).

Inverse temperature compensation has been demonstrated in fishes, reptiles, and amphibians and results in a depression of energy metabolism during torpor in these vertebrates (Aleksiuk, 1976; John-Alder, 1984; Johansen and Lykkeboe, 1979; Walsh et al., 1983; Gatten, 1978; Carey, 1979). There are two related aspects of inverse temperature compensation. One, which probably relates more to those ectotherms at the altitudinal limits of their distribution, involves an increase in the Q_{10} at lower environmental temperatures. Thus the Q_{10} of resting oxygen consumption in *Bufo b. boreus* is 3.72 between 5° and 10°C, but 1.42 between 25° and 30°C (Carey, 1979). This shift represents a strategy for rapidly going into, and coming out of, a metabolic depression on a diurnal basis. This strategy can be compared to the tropical pancake tortoise, *Malacochersus tornieri,* which opts for positive acclimation and demonstrates a Q_{10} of 1–2 (depending upon acclimation temperature) between 20° and 35°C (the temperature range of its environment) (Wood et al., 1978). The other aspect of inverse temperature compensation relates to those vertebrates that spend considerable periods of the year at low temperatures. In these animals, we find that the metabolic rate at a low acclimation temperature is lower than can be accounted for by temperature alone (Walsh et al., 1983). In the cold-acclimated animals, there is an increase in Q_{10} and a downward translation of the whole rate–temperature curve (Figure

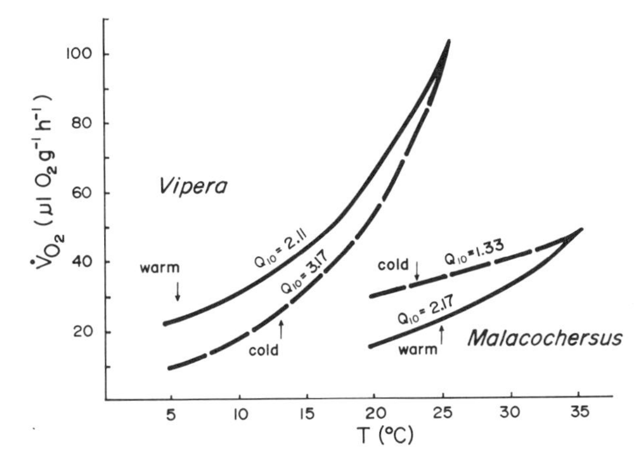

Figure 4.1 Oxygen uptake versus ambient temperature for cold- and warm-acclimated *Vipera berus* (*left*) and *Malacochersus tornieri* (*right*). Acclimation temperatures were 5° and 25°C for *V. berus* and 20° and 35°C for *M. tornieri*. Modified from Johansen and Lykkeboe (1979).

4.1). Again this can be compared to the strategy of *M. tornieri,* in which there is an *upward* translation of the rate–temperature curve and a *decrease* in the Q_{10} in cold-acclimated animals (Figure 4.1). What are the advantages of this seemingly counterproductive strategy, and what are the mechanisms that control it?

Why Go Torpid?

The decision to undergo torpor is obviously an evolutionary one and relates to the costs and benefits of torpor or activity during a stressful period. In those animals that undergo torpor, it may be metabolically feasible to remain active, but the cost is obviously prohibitive. Thus activity and feeding are abandoned and the animal has to rely on *body* fuel reserves during the torpid period. It is thus advantageous (1) to enter torpor quickly, (2) to set torpor metabolism as low as possible, and (3) to get out of torpor as quickly as possible if and when ambient temperature rises—hence, the high Q_{10} values and resultant low metabolic rate at the torpor temperature, and concomitant minimal fuel usage. The quantitative effect of this strategy can be easily demonstrated using the data in Figure 4.1. If a warm-acclimated *V. berus* weighing 50 g uses 20 µl of O_2 g^{-1} hr^{-1} at 5°C it will use up its total fat reserve in seven months (assuming that the animal's fat reserves are 5% of body weight and that the animal is using only fat as fuel). *Vipera berus* in Denmark is in torpor for six to eight months (Johansen and Lykkeboe, 1979). An animal that was not cold adapted would have little or no margin of safety. The same snake, however, if cold adapted, would have fourteen months of fuel supply at 5°C. Fat appears to be the major fuel supply during torpor in most ectotherms, and stores are depleted during this period by 30–60% (Derickson, 1976; Pollock and MacAvoy, 1978; Bennet and Dawson, 1976). It is obvious that if Q_{10} values were not elevated, lipid stores would not last for the entire period of torpor.

Fuel stores are not the only limiting factor during torpor, however. For freshwater turtles, for instance, O_2 becomes as scarce as body fuels. *Chrysema picta bellii* can remain in torpor at 3°C for six months. During this time the animal is submerged, presumably to escape predators and freezing; therefore it has no access to air (although in theory it can meet a small fraction of its metabolic demand for O_2 via aquatic respiration). It is obviously advantageous for this animal to depress metabolism during torpor as much as possible, not only to conserve fuels, but to minimize the potential oxygen debt and associated end-product problems. And it is not surprising to find that the Q_{10} of the metabolic rate for this turtle, between 15° and 5°C, when denied access to air, is a massive 11.0 (Herbert and Jackson, 1985). This strategy is obviously widespread and powerfully effective, possibly providing an insight into the different thermal sensitivities of ectotherm and endotherm cells. Unfortunately, the basis for the increased temperature sensitivity is poorly understood.

Mechanisms

In cold acclimated *V. berus*, the P_{50} and the red cell nucleoside triphosphate concentrations are lower than in warm-acclimated animals. The resulting increase in blood-oxygen affinity could mediate a reduced O_2 diffusion gradient between capillary and cellular sites (Johansen and Lykkeboe, 1979). However, in vitro metabolic rates of heart homogenates from garter snakes acclimated to simulated hibernating conditions are strongly depressed relative to those from individuals acclimated to simulated "cold weather" conditions (Aleksiuk, 1976). The inverse compensation strategy thus may have a physiological and a molecular component, both of which are necessary for an effective inverse compensation. In garter snakes, metabolic depression is seen in heart but not in liver and skeletal muscle homogenates. The depression of the metabolic rate of heart, and therefore of heart rate during hibernation, may depress metabolism in other tissues by reducing blood flow (see Figures 2.1 and 2.2). So blood O_2 affinities and blood flow may be two key physiological parameters contributing to an inverse temperature compensation. At the molecular level there is some evidence that reduction in levels of enzyme activities occurs in cold-adapted animals (Aleksiuk, 1976; Hazel and Prosser, 1970; Walsh et al., 1983). However, in eels, respiratory adjustments as a result of lowering 10°C- or 15°C-acclimated animals to 5°C, take place in 1 hr and may be related to more rapid phenomena such as changes in pH with temperature (Walsh et al., 1983). A completely different, yet hitherto unresolved area, involves inhibitory subtances. There is evidence that taurine and GABA can cause depression of neuron activity and that levels of both increase in the brain during hibernation in the lizard *Varanus griseus* (Raheem, 1980; Raheem and Hanke, 1980). So, although this phenomenon has great potential in fields such as hypothermic surgery, we really do not understand the mechanisms that control it, and the area must surely be one in which more research effort would be profitable.

Sliding down a Q_{10} curve has its advantages, but there are problems that are encountered on the way and must be overcome. Two of these problems are acid–base status and ion homeostasis.

Acid–Base Concepts in Ectotherms

The pH concept was introduced as an easy way to describe the acid–base status of homeotherms and is very useful in that context. In ectotherms, however, the pH concept alone is inadequate, and other parameters are necessary to describe in meaningful terms the acid–base status of the animals.

Water at neutrality provides an excellent example of a solvent that maintains an equal dissociation between H^+ and OH^-, even though its pH decreases as the temperature rises. Thus, as the temperature rises from 5° to 30°C, the pH at which $[H^+] = [OH^-]$ (pN) decreases from 7.37 to 6.92 (Howell and Rahn, 1976) (Figure 4.2). Ectotherm blood in vitro (sealed to maintain a constant total CO_2 content) is more alkaline than distilled water at any temperature, but the rate of change of pH per degree Celsius is often the same (Figure 4.2).

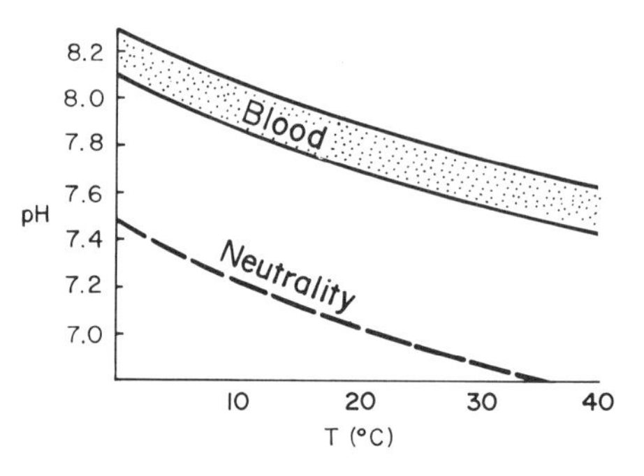

Figure 4.2 The top band represents blood pH values from 36 species at the given temperatures. The lower line is pN. Modified from Howell and Rahn (1976).

Sealed, in vitro blood maintains a constant OH^-:H^+ ratio as temperture changes, *and,* more to the point, so does in vivo blood from a wide variety of ectotherms. Three pertinent questions arise:

1. What property of the blood produces the same ΔpH/°C as that of water?
2. What is the advantage to the animal of this pattern of acid-base behavior with temperature?
3. By what mechanism do ectotherms maintain a constant CO_2 content in the blood as temperature changes, a mechanism allowing the above pattern to emerge in vivo?

Imidazole groups in proteins. Bicarbonate, phosphate, and protein constitute the three major buffers of blood. The buffering component of protein is the imidazole group of the amino acid histidine. The change of the pK of imidazole with temperature is the same as the change in pN with temperature. The imidazole groups are thus the cause of the observed relationship between the pH of blood and the pN (Howell and Rahn, 1976).

Why defend the ΔpH/°C of imidazole? Enzymes are very sensitive to titration of their histidine imidazole groups. Thus it is an advantage, in terms of conservation of regulatory potential, to maintain at all temperatures a constant value of the dissociation ratio of imidazole groups (α_{Im}) (Malan, 1983).

How do ectotherms defend the α_{Im}? This final question leads us into a discussion of the physiological significance of the pH–temperature concept and enables us to relate pH, temperature, and the high Q_{10}

values discussed earlier. The environment and the torpor strategy determine how this problem is solved, so we shall discuss this question from three points of view: that of the air-breathers, that of the water-breathers, and that of the breath-holders.

Air-Breathers

The air-breathers defend the α_{Im} by maintaining a constant total CO_2 content. This is accomplished by varying the air convection requirement (ACR), which is the ratio of respiratory minute volume to metabolic rate. The ACR decreases with temperature in most air-breathing ectotherms in which it has been measured. The physiological control of this ratio may be via a receptor for [HCO_3^-] or via one that includes a titratable imidazole group (Jackson, 1978; Reeves, 1977).

As a consequence of this strategy, an animal with a low extracellular fluid [HCO_3^-] would require a higher ACR than an animal with a higher ECF [HCO_3^-]. The [HCO_3^-]—and this is species-specific—thus dictates what both the P_{CO_2} and the ACR must be in order to achieve the proper pH at any temperature. But ACR also affects the supply of oxygen; so a relatively high ECF [HCO_3^-] would benefit animals preferring a low temperature range (Jackson, 1978). The [HCO_3^-] varies among families of reptiles and among reptiles in each family (Gregory, 1982), but there is no correlation between [HCO_3^-] and minimum environmental temperature encountered by the animal. What of changes in [HCO_3^-] with season in those air-breathing ectotherms that encounter low winter temperatures? The data on seasonal [HCO_3^-] are sparse, but interesting. Bicarbonate concentration, instead of rising during the winter—a change that would enable the animal to maintain the "proper" pH at a lower ACR (Jackson, 1978)—*drops* during winter (Gregory, 1982; Haggag et al., 1965). The inevitable result of this would be that CO_2 content would not remain constant; thus the ΔpH/°C would not be conducive to maintaining α_{Im}. Vertebrate ectotherms do not always defend α_{Im}. Two notable exceptions are the varanids *gouldii* and *exanthermaticus,* which do not experience very low temperatures but do defend a constant pH between 25° and 35°C (Gregory, 1982). So although the general trend in air-breathing ectotherms is to defend α_{Im} at the expense of pH, there are exceptions. Unfortunately, there are no data on good examples of ectotherm hibernators, specifically at temperatures between 5° and 15°C. We therefore tentatively suggest that a deviation from the alphastat pattern, by air-breathing ectotherms as they go into torpor, is a positive mechanism resulting in the lowering of the efficiency of metabolism. This hypothesis blends the physiological and biochemical data and explains the high Q_{10} values that are measured at low temperatures.

Water-Breathers

In fishes, blood P_{CO_2} is determined by environmental conditions. Thus fishes do not maintain a constant CO_2 content, but resort to slower mechanisms such as gill ion-exchange processes to change the bicarbonate concentration of plasma. Although there is a general trend for the blood to follow the alphastat model, again there are exceptions (Reeves, 1977; Heisler, 1980). Another complication in fishes, because the whole process is slower and involves ion pumping, is that the pattern of pH regulation with temperature can be organ-specific (Walsh and Moon, 1982). This characteristic may be a useful adaptation that is denied to the air-breathers (Chapter 5). The fishes therefore can overcome the limitation of their aqueous environment as far as pH regulation is concerned. As in the air-breathers, the exceptions may be those fishes in which Q_{10} is indirectly increased at low temperatures through unfavorable effects of temperature on pH.

Breath-Holders

The turtles create for themselves a special problem during torpor. Their blood behaves in the "normal" way in vitro and in vivo as long as they are breathing air (Howell and Rahn, 1976; Reeves, 1977). However, animals such as the painted turtles of North America choose to hibernate not just *under* the water but in *anoxic* water at about 3°C (Ultsch et al., 1984; Jackson et al., 1984). This strategy is advantageous in that it increases predator avoidance and decreases metabolic rate both through a temperature effect *and* presumably by inhibiting the Pasteur effect (Chapter 2). The decrease in total metabolic rate, between air-breathing at 20°C and breath-holding in anoxic water at 3°C, is 160-fold; at 3°C, normoxic–anoxic transition leads to a 50- to 60-fold metabolic suppression. These arrests minimize the need for fermentation reactions, with their concomitant complications (Figure 4.3). Lactate production does occur, however; and the location of the torpid turtle prevents it from using the two previously discussed methods of regulating pH as temperature changes. The turtle is not ventilating and thus cannot alter its ventilation rate and, because aquatic respiration is limited, exchange of HCO_3^- and CO_2 is also

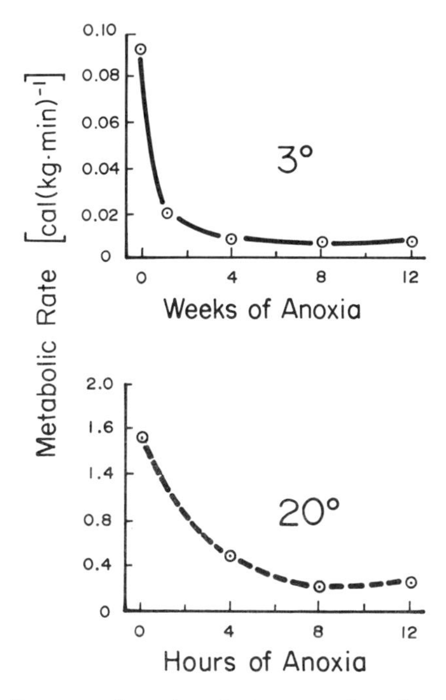

Figure 4.3 Metabolic rate of turtles during anoxic submergence at 3° and 20°C. The rate at time 0 is the measured aerobic rate, and the subsequent values are estimated from plasma lactate accumulation. Modified from Herbert and Jackson (1985).

limited. So lactate is produced, a reaction that causes a metabolic acidosis. The resulting titration of HCO_3^- causes a respiratory acidosis that is compounded by the limited capacity of the torpid turtle to exchange CO_2. At first glance, it would seem that torpid turtles face an insoluble dilemma, but recent experiments have shown what we should have by now come to expect—the metabolic perturbations not only are tolerated but are put to good use.

The root of the torpid turtles problem is lactate production, so those turtles that can more effectively minimize lactate production (that is, depress metabolic rate during submergence) have the greatest tolerance to anoxia. The turtles with the greatest tolerance to anoxia, *Chrysemys* and *Chelydra,* show changes in plasma [HCO_3^-], plasma [lactate], blood pH, and P_{CO_2} during submergence quantitatively different from those shown by *Trionyx* and *Sternothermus,* which are poorly adapted to anoxia (Table 4.1). Some of these differences may at first seem contrary to those expected, but it all becomes clear if CO_2 retention is considered to be a metabolic depressant. *Chrysemys* and

Chelydra are poorly adapted with respect to extrapulmonary CO_2 loss; thus they accumulate CO_2, a process that causes a respiratory acidosis and its accompanying lowered pH; in this way metabolism and thus lactate accumulation is inhibited (Table 4.1). *Trionyx* and *Sternothermus,* on the other hand, are better adapted with respect to extrapulmonary CO_2 loss; they minimize hypercapnic acidosis but promote more rapid anaerobic metabolism and thus metabolic acidosis. Indeed, the acidosis in *Trionyx* and *Sternothermus* is predominantly due to accumulated lactate, whereas in *Chrysemys* and *Chelydra* it is predominantly due to elevated P_{CO_2} (Jackson et al., 1984; Ultsch et al., 1984).

As is so often the case, there is a continuum in strategy that is evident from the fishes, through the lizards, snakes, and frogs, to the turtles. The effect of temperature on pH is to some extent modified, but also utilized to generate high Q_{10} values and thus low metabolic rates during torpor. Fuel and oxygen conservation are thus ensured, and anaerobic end products are kept at low levels.

Ion Homeostasis

Hypothermia is lethal in nonhibernating mammals and is characterized by the failure of the ability to rewarm and ultimately by cardiac arrest at temperatures between 13° and 20°C (Angelakos et al., 1969). Although hypothermia affects the cell's biochemistry at various levels (Stoner et al., 1983), the cause of cell death at low temperatures is

Table 4.1 Effect of submergence on various parameters in turtles

Parameter	*Chrysemys*	*Chelydra*	*Sternothermus*	*Trionyx*
Control Plasma [HCO_3^-] (m*M*)	36	39	27	24
Plasma [lactate] after 50-hr submergence	25	20	40	35
Minimum pH during submergence	6.8	6.8	7.0	7.0
Maximal decrease (%) in plasma [HCO_3^-] during submergence	65	63	78	80
Maximal pCO_2 (mm Hg) during submergence	55	70	27	23

Source: From Ultsch et al. (1984).

almost certainly located at the level of the cell membrane. Membrane ion-transport processes can account for 45% of the resting metabolism of a cell. They are perturbed by changes in temperature and have been extensively studied in relation to temperature and cell death (Hazel, 1973; Whittam et al., 1964; Glitsch and Pusch, 1984).

Temperature affects the Na^+,K^+-ATPase pumping system through the enzymes involved and through membrane permeability per se (Winter, 1973). Problems of ion imbalance arise at low temperatures because the enzyme pump and membrane permeability, or leak, are affected differentially by temperature. Ion imbalance—loss of K^+, to be precise—is the only consistent, potentially damaging result of cold exposure seen in mammalian cells. It is therefore strongly implicated as a major factor involved in cold death (Willis, 1979; Willis et al., 1980; Lyman et al., 1982).

The effect of low temperature sets off an ion-related chain reaction. Potassium ions in the ICF and ECF shift toward their equilibrium concentrations (Figure 4.4) (Willis and Baudysova, 1977). Sodium ions move in the opposite way; and if this process is not interrupted, it ultimately leads to partial membrane depolarization, the opening of voltage-dependent Ca^{++} channels, and the influx of Ca^{++} (Hansen, 1982). An example of the effect of increasing intracellular Ca^{++} concentrations is shown in the following reactions.

$$\text{phosphatidylcholine (PC)} \xrightarrow{H_2O} \text{lysoPC} + \text{arachidonic acid}$$

$$\text{phosphatidylethanolamine (PE)} \xrightarrow{H_2O} \text{lysoPE} + \text{arachidonic acid}$$

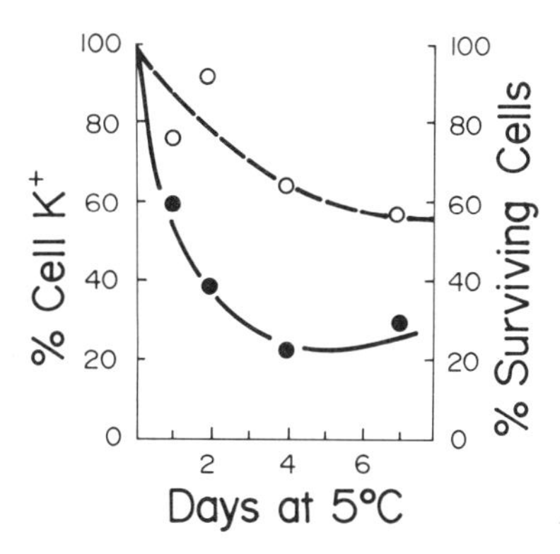

Figure 4.4 Retention of K^+ and survival of cells at 5°C from human embryonic lung. ○, cell K^+ (percentage of original cell K^+ before exposure to cold); ●, percentage of cells still living. Modified from Willis and Baudysova (1977).

The end products of these reactions normally are formed after thrombin activation of platelets, but during hypothermia they proceed without thrombin, a result indicating a low-temperature-mediated "activation" of phospholipase A_2. From parallel studies it is known that the influx of Ca^{++} occurs prior to the onset of lysoPC and lysoPE formation and the liberation of arachidonic acid. Thus either thrombin activation or hypothermia is able to initiate a similar chain of metabolic events. The difference between the two pathways is that thrombin activation is a controlled process whereas the hypothermia-mediated process is apparently largely uncontrollable. The pathogenic pathway favored by prolonged hypothermia may be autocatalytic in the sense that cell membrane damage may in turn accelerate further dissipation of ion gradients.

Vertebrate ectotherms apparently deal with this potential obstacle at its very roots: ion gradients are obviously highly conserved parameters and significant perturbations of them cannot be tolerated, even by these masters of improvization. Thus changes in plasma ion concentrations are nonexistent or slight. Plasma Na^+, K^+, and Ca^{++} levels are not altered during short-term (hours) temperature changes (Munday and Blane, 1961; Bickler, 1984), and even over a full torpor period of six months the maximum change observed is 25% (Haggag et al., 1965; Gregory, 1982). The situation differs somewhat in the torpid turtles who have to try to compensate for lactate production through ion adjustments (Jackson et al., 1984). The extracellular $[K^+]$ in the brain of a 4-hr anoxic turtle only shows a moderate rise compared with the 1600% rise in a 30-min, anoxic rat brain (Sick et al., 1982).

Again we are observing a crucial adaptation that probably more than any one other process embodies the nature of cold tolerance. Unfortunately, we can only offer data that merely suggests the underlying mechanism.

Leaky and Nonleaky Membranes

This question has already been considered in Chapter 2, a discussion culminating in the channel arrest hypothesis. We will reconsider it only briefly here in so far as it relates to the ectothermic vertebrates.

Ion channels are specific proteins and can be regulated by controlling the rate and voltage-dependence of opening and closing (Catterall, 1984). As well as regulating ion flow in this manner, regulation can be on a tissue-specific basis (some tissues have higher densities of channels than others; Catterall, 1984), or on a time basis (by varying the density in a tissue between different steady states; Li et al., 1982; Palmer et al., 1982).

We suggest that the same mechanism and strategy may explain the observed (Hulbert and Else, 1981) permeability differences between ectotherms and endotherms. Although systematic, channel-density studies of homologous ecto- and endothermic tissues designed to quantitatively test this idea are not available, it is notable that in the rabbit Na^+ channel densities in unmyelinated nerves (vagus) average about 110 pores per square micrometer, whereas in analogous axons in ectotherms (lobster and garfish), channel densities are 90 and 35 per square micrometer, respectively (Catterall, 1984). Similarly, in rat fast, slow, and diaphragm muscle, Na^+ channel densities of 557, 371, and 421 per square micrometer, respectively, contrast with densities in frog skeletal muscle membranes as low as $195/\mu m^2$ (Rogart, 1981). The occurrence of lower Na^+ channel densities not coincidentally correlates with Na^+,K^+-ATPase activities (Na^+ pump densities) in cell membranes from ectotherms that are lower than those in endotherms (Hulbert and Else, 1981; Edelman, 1976; Stevens, 1973; Stevens and Kido, 1974).

It is possible that hypothermic cell death due to ion imbalance is at least partially offset by minimizing the offending arm of the process—the "leak." However, one cannot help but think that this is not the only strategy involved. As mentioned previously, many ectotherms spend eight months below 10°C, in which case the leak would have to be completely plugged, or counteracted, to prevent lethal imbalance. So apart from the quantitative solution mentioned earlier, is there any other complementary mechanism that could offer a more permanent solution and help to plug the remainder of the leak? The Q_{10} of both glycolysis and the pump per se are likely contenders for the focus of this additional strategy.

Tailoring the Q_{10} of Glycolysis and the Na^+ Pump

Apparently glycolysis is more effective than oxidative phosphorylation in supporting Na^+,K^+-ATPase (Balaban and Bader, 1984; Kutchai and Geddis, 1984). One explanation for this linkage is that the protons produced during glycolysis are extruded from the cell in exchange for Na^+, which is then exchanged for K^+ (Balaban and Bader, 1984). Another explanation is that the Na^+,K^+-ATPase pump provides ADP, which is limiting for glycolysis (Kutchai and Geddis, 1984).

If glycolysis does supply the ATP for the Na^+, K^+-ATPase pump, the effect of temperature on the pump may not be related only to the Q_{10} of the pump per se. If ATP is limiting, it may also involve the Q_{10} of glycolysis. Thus an ectotherm could tailor both the Q_{10} of the pump and that of glycolysis in order to minimize the differential effect of temperature on pump and leak. Ion imbalance would not occur and the Q_{10} of the major metabolic pathway supporting the bulk of cell maintenance during torpor (β-oxidation) need not be similarly affected. There are two lines of evidence for this scheme. First, ischemia causes loss of ion homeostasis within 60–120 min in the turtle, in contrast to anoxia, which takes up to 60 hr. This finding suggests that substrate for an anaerobic metabolic pathway, perhaps glycolysis, is necessary for ion homeostasis (Sick et al., 1985). Second, the Q_{10} of the sodium pump does vary between mammalian species and with temperature (Willis et al., 1980; Glitsch and Pusch, 1984), but the effect of temperature on the sodium pump in ectotherms has not been measured. The turnover rates of both glucose and palmitate in the bobtail lizard *Tiliqua rugosa,* which winters at about 15°C, show increased temperature sensitivity at low temperature ranges. The Q_{10} of the turnover rate of palmitate is higher than that of glucose at all temperatures, however (M. Guppy, unpublished data). These data support the hypothesis that the Q_{10} of β-oxidation is independently adjusted to suit an effective inverse temperature compensation strategy.

Low-temperature torpor is a complicated process and a veritable biological obstacle course. But at least for the vertebrate ectotherms it is in a way an extension of their normal daily regime. These animals are already designed to endure variable body temperatures, and they have only to stretch their strategies to cover longer intervals at lower temperatures. What of those animals, however, that have a high (37°C) and virtually invariable "normal" T_b but also hibernate, for months at a time, with a T_b of 5°C? These are the mammalian hibernators and we turn our attention to them in the next chapter.

Summary

Many vertebrate ectotherms encounter low ambient temperatures for months at a time. Instead of attempting to counteract the effect of a declining temperature on metabolic processes, these animals exaggerate the effect by raising Q_{10} values at low temperatures. The result is a very low torpor metabolism, which can be fueled by endogenous fat stores and in some cases minimizes the need for anaerobic pathways during anoxia. The mechanisms proposed for this inverse temperature compensation range from organ–tissue interaction, through Hb–O_2 affinity, to specific neuroinhibitors. The problems encountered as a result of accepting both T_b changes and reduced metabolic rate relate to pH and ion homeostasis. The pH necessarily changes with temperature and profoundly affects enzyme function. The air- and water-breathing vertebrate ectotherms are able to perturb the pH–temperature relationship to their advantage and may use it to produce the large Q_{10} values at low temperatures. The breath-holding vertebrate ectotherms have the additional problem of lactate acidosis and can encourage respiratory acidosis as a feedback mechanism to slow metabolic rate and thus to forestall metabolic acidosis. Ion homeostasis raises its head every time metabolic arrest is encountered. The channel arrest hypothesis may well be the answer here, as it is in anaerobes and mammalian divers. Tailoring the temperature sensitivity of specific Na^+,K^+-ATPase pump-linked pathways may be another way that these animals prevent ion imbalance during the metabolically arrested state at low temperatures. The time extension factor in this case refers to fuel supplies and anaerobic end products. One gram of fat in a cold-adapted lizard is equivalent to 2–5 g of fat in a nonadapted lizard. The lactate levels that result from six months of torpor in *Chrysema picta bellii* would normally appear after only four days of anoxia in a nonadapted animal.

5

Endothermic Hibernators

The ability to hibernate is widespread in the mammals. Animals that can temporarily lower their body temperature, heart rate, and metabolic rate are found in the three mammalian subclasses and in five orders of placental mammals (Lyman et al., 1982). The three orders that exhibit the more profund and most regular torpidity are the rodents, the bears and the bats. We shall restrict our discussion to rodents and bears to demonstrate the adaptations that are seemingly requisite for mammalian hibernation.

Rodents

Three genera of rodents (ground squirrels, marmots, and chipmunks) regularly hibernate, but only the ground squirrels and the marmots exhibit deep hibernation, during which the body temperature falls below 10°C, the heart rate and metabolic rate are greatly reduced, and the animal maintains the ability to rewarm using only endogenously generated heat (Lyman et al., 1982). The ground squirrels (genus *Citellus/Spermophilus*) are the most popular biological material for hibernation studies and have provided the most data. The duration of the hibernation period in the ground squirrels varies with the geographical location, but is maximally eight months (Davis, 1976; Mitchener, 1977). Periodic arousal is a characteristic of these deep hibernators (Lyman et al., 1982), and the duration of the actual torpid period in the ground squirrels ranges from 80 to 150 days per year, with maximal uninterrupted torpid periods of one month (Wang, 1978; Pengelley and Fisher, 1961). Ground squirrels do occasionally store food (Davis, 1976), but either they do not eat during arousal

(Wang, 1978; Pengelley and Fisher, 1961) or food intake is very low, being about 5% of prehibernating levels on a per day basis (Jameson, 1965). Weight loss during the hibernation period is 40–50% of body weight (Galster and Morrison, 1976), most of this occurring during the homeothermic periods (Jameson, 1965). During torpor, body temperature is between 2° and 10°C and is usually within 4°C of ambient temperature (Lyman et al., 1982; Wang, 1978), O_2 consumption is 4–10% of normal (Wang, 1978; Hammel et al., 1973), the respiratory quotient (RQ) is 0.78 (Snapp and Heller, 1981), and the heart rate drops from 500 to 20 beats/min (Lyman et al., 1982).

The data on the marmot genus indicate the same changes from active to dormant states (Lyman et al., 1982; Davis, 1976; Bailey and Davis, 1965; Smith and Hock, 1963; Lyman et al., 1958), although the prehibernating heart rate is about 100 beats/min and drops to 8–10 beats/min during hibernation (Lyman et al., 1958).

The chipmunks differ considerably from the marmots and the ground squirrels and are not really deep hibernators. They are erratic hibernators; their hibernation period is shorter and the periods of torpor last no longer than six days (Davis, 1976; Wang and Hudson, 1971). The chipmunks store food and eat, apparently voraciously, between torpor bouts (Davis, 1976, Neumann, 1967). Body temperature during torpor falls to a minimum of 20°C, O_2 consumption drops by 66–98%, and heart rate decreases from 165 beats/min to 23–46 beats/min (Wang and Hudson, 1971). Ground squirrels have been used for the majority of studies on hibernation, and most of the data below concerns these animals.

The metabolism of a hibernation bout can be partitioned into three phases. The first is entry, which takes about 20 hr; the second is torpor, which lasts from a few days to a month; and the third is arousal, which lasts about 2 hr and grades into homeothermy (Morrison and Galster, 1975; Wang, 1978; Snapp and Heller, 1981).

Entry

Entry into torpor accounts for about 13% of the energetic cost of hibernation (Wang, 1978). It is characterized by decreasing heart rate, breathing rate, T_b, blood pressure, and O_2 uptake (Figure 5.1). The decreasing heart rate is certainly due to parasympathetic influences (Lyman et al., 1982), and the decline in T_b, characterized by increasing heat loss and decreasing heat production, is actively facilitated, probably through vasodilation (Wang, 1978; Lyman et al., 1982; Hammel

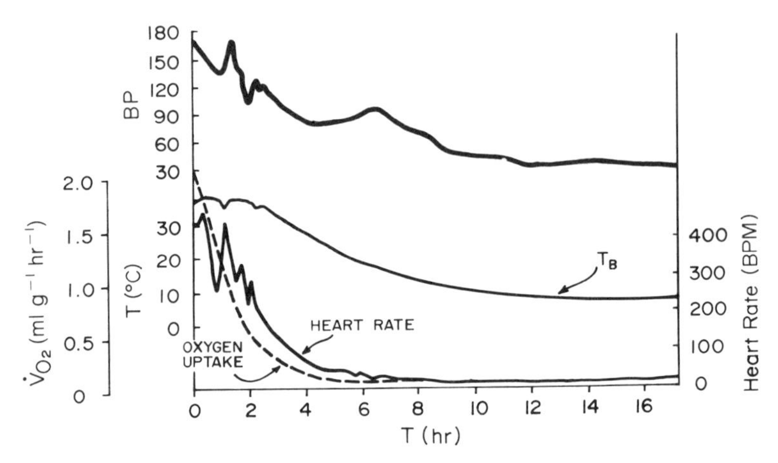

Figure 5.1 Blood pressure (BP), temperature (T°C), and heart rate of a thirteen-lined ground squirrel entering hibernation (after Lyman, 1965). The oxygen uptake ($\dot{V}_{O_2}$) is from a Richardson's ground squirrel (Wang, 1978).

et al., 1973; Tucker, 1965; Heller et al., 1977). The ability to thermoregulate, however, is not lost during torpor in the ground squirrels. The temperature set point, as determined by the hypothalamus, gradually decreases with each successive "test" entry into torpor until a minimum is reached during the entry preceding a long torpor bout. At all times, however, metabolic rate is sensitive to hypothalamic temperature. Warm-sensitive neurons in the hibernator are therefore necessarily sensitive at unusually low temperatures. Warm-sensitive neurons in the golden hamster display a continuous firing rate down to a hypothalamic temperature of 14°C; those of the guinea pig are silent below 28°C (Heller et al., 1978). In contrast to heart rate and temperature changes, the mechanisms underlying the severe metabolic depression, and their relation to T_b changes, are poorly understood.

Suppression of Metabolism

The active suppression of metabolism has been the subject of much investigation, but with few conclusive results. Abnormally high Q_{10} values have been suggested as a mechanism, but recent rigorous experiments have shown that the Q_{10} of O_2 uptake during entry, once the thermogenic factor is discounted, is within normal limits, between 2.0 and 2.5 (Snapp and Heller, 1981). These figures increase to about 8 if the thermogenic factor is included, so the removal of the latter accounts for much of the observed metabolic suppression. Whole-body temperature sensitivities are not therefore abnormally high, but perturbation of Q_{10} values can play a role in a tissue-specific sense.

In awake, heterothermic ground squirrels, blood acid–base balance is similar to that of the vertebrate ectotherms (Bickler, 1984). However, rodents entering torpor develop a respiratory acidosis that is reflected by a decrease in RQ and is equivalent to about 0.3 pH units (Lyman et al., 1982; Snapp and Heller, 1981). If the respiratory acidosis is compensated—by the metabolic extrusion of acid in return for ions (Lyman et al., 1982)—the pH ($[H^+]$) varies with temperature, but the $[OH^-]:[H^+]$ ratio remains constant (that is, the blood pH follows the alphastat model; see Chapter 4). If the respiratory acidosis is uncompensated, pH ($[H^+]$) is constant with temperature, but the $[OH^-]:[H^+]$ ratio varies. The "choice" made by a tissue—whether to compensate or not as temperature drops—has a profound effect on the effect of temperature on the catalytic rate of its enzymes.

The values for the $K_{m(\text{substrate})}$ of many enzymes in vitro vary with temperature if the pH of the reaction is held constant (uncompensated). The affinity of the enzyme for its substrate increases as temperature drops, and the Q_{10} of the reaction is consequently lowered (Figure 5.2). The effect of temperature upon reaction velocity, however, is strongly dependent on pH (Wilson, 1977). If the pH of the reaction mixture is varied to hold the $[OH^-]:[H^+]$ ratio constant (compensated), the K_m changes are less and consequently the Q_{10} values for the reaction are higher (Figures 5.2 and 5.3).

The blood and brain of the ground squirrel are completely uncompensated and the muscle is only slightly compensated (Malan et al., 1981). The metabolism of these tissues would therefore be subject to Q_{10} values that are lower than those in the heart and the liver, where complete compensation occurs (Malan et al., 1981). Thus, in the heart and the liver of the rodent hibernator, the high Q_{10} values would effectively turn down metabolism as T_b drops during entry. In muscle, blood, and brain, where Q_{10} values are relatively low, respiratory acidosis—well known as an effector of cellular processes (Lyman et al., 1982)—would be the major suppressive factor. In keeping with this suggestion, PFK from the skeletal muscle of *C. Beecheyi* is much more prone to pH regulation at 6°C than it is at 37°C. In fact, at low

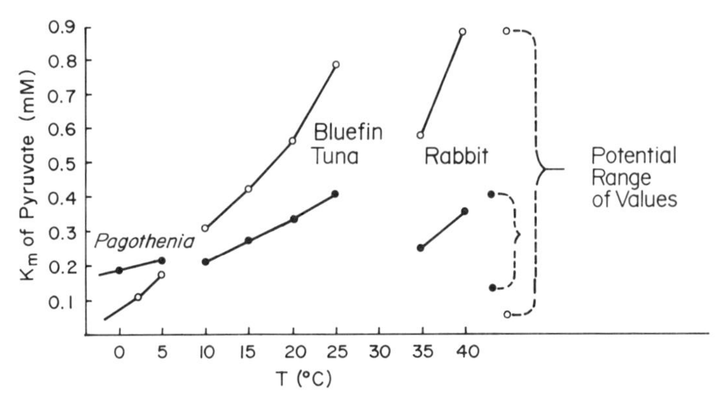

Figure 5.2 The effect of temperature on the apparent K_m of pyruvate for M_4 lactate dehydrogenases (LDHs) of vertebrate species adapted to different temperatures. Assays were performed under two pH regimes: temperature-dependent pH in an imidazole buffer (*closed symbols*) and a constant pH (phosphate) buffer system (*open symbols*). For the latter, the pH was at pH 7.4 at all temperatures. The vertical range to the right of the figure illustrates the spread of K_m values at biological temperatures found under the two assay systems. Modified from Hochachka and Somero (1984).

temperatures, an acidification of 0.3–0.4 pH units causes a 60% deactivation of PFK (Hand and Somero, 1983).

Why are the mechanisms that suppress metabolism as the temperature drops tissue-specific? The answer may be related to differing needs of tissues and organs. It is easy to envisage a situation where a

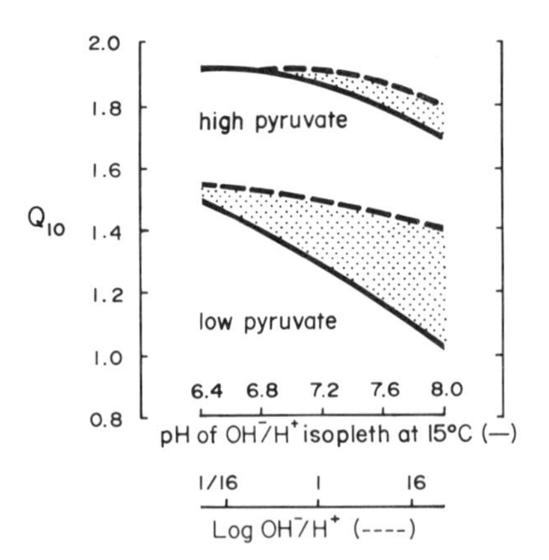

Figure 5.3 Q_{10} values for pyruvate reduction (average values for 5°–15°C and 15°–25°C) compared for constant pH conditions (*dashed lines*) and constant $[OH^-]/[H^+]$ (*solid lines*). After Wilson, 1977.

peripheral muscle is hypothermic when the animal is not hibernating. A high Q_{10} value would render this muscle inactive, but a low Q_{10} value, in the absence of respiratory acidosis, would allow the muscle to continue to function. The same argument can be applied to blood cells that would be perfusing tissues at different temperatures. The same cannot be said for the brain, however, which after all is completely uncompensated; but acidosis may function in the brain as an inhibitor of thermoregulatory neurons (Malan et al., 1981). Because the heart and liver *never* encounter a hypothermic situation except during torpor, high Q_{10} values constitute an adequate mechanism for metabolic suppression, *and* the acid–base state of these two central organs is maintained.

We saw a similar situation in the fishes and the turtles (Chapter 4). The pH of some fishes appears to be organ-specific, but the rationale devised for the ground squirrels cannot apply, because temperature throughout fishes is homogeneous at all times. The turtles also have a respiratory acidosis, at least during the early part of torpor; but the situation is complicated by the production of lactate and H^+. It is therefore impossible to classify the different species as either compensated or uncompensated as in squirrels.

Qualitative Aspect of Torpor

The torpid period has attracted much interest, presumably because the temperatures and duration involved are so extreme. Qualitative metabolic differences are apparent between the brains of torpid and euthermic rodents. This is manifest in the uptake of 2-deoxyglucose, which either is increased or decreased in parts of the brain in torpid animals. These results are particularly significant in light of the usual reliability of the euthermic pattern. The physiological meaning of most of the differences is unknown, but the paratrigeminal nucleus, which increases its relative uptake of 2-deoxyglucose during torpor, may process thermal information from the head (Kilduff et al., 1982; George et al., 1982; Kilduff et al., 1983). The qualitative aspect of whole-body metabolism, however, offers nothing unexpected.

Lipid. As in the air-breathing ectotherms, fat metabolism is certainly implicated as the main energy supply during torpor. The percentage of body weight accounted for by fat stores doubles between summer and autumn and RQ values drop during torpor (Lyman et al., 1982; Galster and Morrison, 1976; Snapp and Heller, 1981; Bailey and Davis, 1965). Serum lipids rise in late autumn and drop during hibernation; blood albumin levels are highest during torpor (Galster and

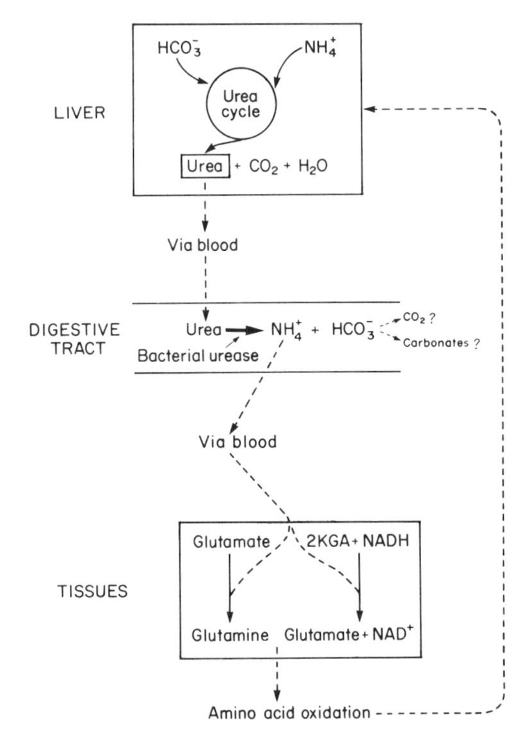

Figure 5.4 Urea cycling between tissues and organs.

Morrison, 1966); but, as would be predicted, actual activities of lipid oxidation and synthesis drop during torpor (Lyman et al., 1982).

Carbohydrate. Glycogen stores and plasma glucose levels decrease during torpor in some species (Galster and Morrison, 1975) but do not change in others (Zimmerman, 1982). There is evidence that the gluconeogenic pathway is actively maintained during torpor (Lyman et al., 1982; Klain and Whitten, 1968), and alanine and glycerol are implicated as substrates (Whitten and Klain, 1968; Galster and Morrison, 1975).

Protein. The rates of protein synthesis and degradation are low in torpid animals (Whitten and Klain, 1968). Plasma urea concentration does not rise markedly during torpor and often drops, and it is not correlated with the length of the torpor bout (Galster and Morrison, 1975; Passmore et al., 1975; Zimmerman, 1982). Renal function virtually ceases during torpor (Deavers and Mussachia, 1980), and the ground squirrel is capable of cycling urea through the gut (Reidesel and Steffen, 1980; Figure 5.4).

Quantitative Aspect of Torpor

While the qualitative analysis of the torpid period is informative, far greater insight arises from assessing quantitative changes that accompany torpor. The O_2 uptake of a resting ground squirrel (250–350 g) is about 250 ml/hr (Wang, 1978; Morrison and Galster, 1975; Galster and Morrison, 1975) and drops to about 10 ml/hr during torpor, or to 4% of normal. If the fuel consumption of a 70-kg man is taken as a standard (Lehninger, 1975) and if the assumption is made that fuel utilization and urea production are proportional to oxygen uptake, the metabolism of a torpid ground squirrel can be quantified (Table 5.1). We can calculate from Table 5.1 that during a 30-day torpor bout, which is unusually long (Wang, 1978; Morrison and Galster, 1975), the animal would use 1.4 g of fat, 4.3 g of carbohydrate, 1.4 g of protein, and produce 0.4 g of urea.

Lipid and protein. Triglyceride contributes 65 g toward the weight of a 300-g ground squirrel in summer, and another 145 g of triglyceride are stored before hibernation begins (Galster and Morrison, 1976). A 300-g ground squirrel contains 57 g of protein (Galster and Morrison, 1976), thus a 30-day torpor bout would only deplete fat and protein stores by 0.7 and 2.4% respectively. Fat and protein reserves are therefore definitely not limiting during individual torpor bouts and *torpor alone* would only deplete reserves of fat and protein by 3 and 12% respectively over the whole hibernation season, assuming a *total* of 150 days of actual torpor.

Carbohydrate and urea. The quantitative changes are obviously highly effective when applied to fat and protein stores, but even this drastic 97% decrease in metabolism does not solve the two problems of fuel supplies for the brain and urea production. The mammalian

Table 5.1 Oxygen consumption and fuel utilization

Subject	O_2 consumption (ml/hr)	Fat util. (g/hr)	CHO util. (g/hr)	Protein util. (g/hr)	Urea prod. (g/hr)
70-kg man	14,800	2.5	8.3	2.9	1.0
Active 300-g ground squirrel	250	0.04	0.14	0.05	0.02
Torpid 300-g ground squirrel	9	0.002	0.006	0.002	0.0006

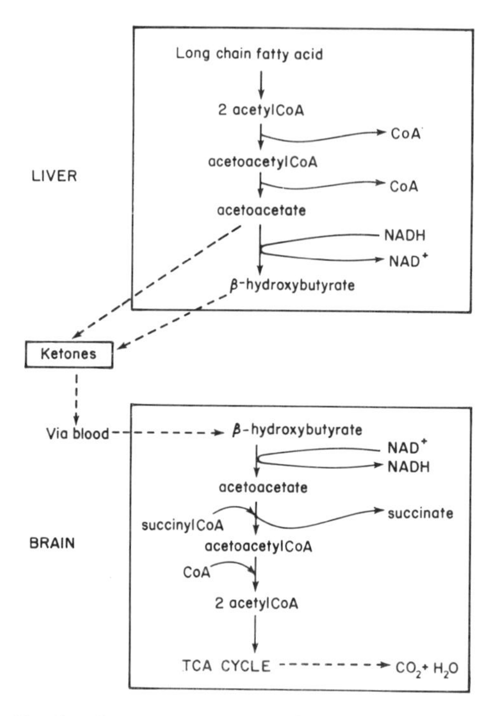

Figure 5.5 Production, transport, and oxidation of ketone bodies.

central nervous tissue cannot utilize fatty acids as an energy source (Allweis et al., 1966), and glycogen stores in the rat liver are 99% depleted after 36 hr of starvation (Newsholme and Start, 1973). The liver of the ground squirrel contains about 0.2 g of glycogen (Galster and Morrison, 1975; Zimmerman, 1982), assuming liver to be 4% of body weight, an amount that could not sustain a single torpor bout, let alone 150 days of torpor (Table 5.1). How could the carbohydrate deficit be made up? It could come from the glycerol released during triglyceride lipase activity, from proteins or from ketones. The first two processes, for a single 30-day bout, would require 41 g of triglyceride, or 7.4 g of protein (Lehninger, 1975, p. 842; Newsholme and Start, 1973, p. 257); thus, 150 days of torpor would consume virtually all of the lipid and protein reserves. An animal could not survive such a depletion of body protein, and most of the ground squirrel's fat reserves are needed for other periods of the hibernation cycle. The inescapable conclusion is that ketones must make up the deficit (which would require only 2.7 g of fat per 30-day torpor; Figure 5.5

and Table 5.2) and, if glycogen levels do fall during torpor, they are restored between torpor bouts.

Urea production during torpor, extrapolating from the 70-kg man (Table 5.1), is 0.43 g per 30-day torpor. If this were added to the extracellular fluid of a ground squirrel (assuming extracellular fluid to be 19% of body weight), the concentration of urea in the blood would be 120 m*M*, which is ten times the normal mammalian plasma urea concentration (Lehninger, 1975). Plasma urea concentrations are never above 10 m*M* in the ground squirrel (Galster and Morrison, 1975; Passmore et al., 1975), so either protein metabolism is reduced beyond that indicated in Table 5.1, a conjecture suggested by the data of Galster and Morrison (1976), or, because of urea cycling, the nitrogen produced does not accumulate as urea, or both.

Urea cycling. Urea cycling has been demonstrated in ground squirrels (Reidesel and Steffen, 1980), and it behooves us to consider the

Table 5.2 Calculated carbon budget for a 300-g ground squirrel for a 6-month hibernation period, assuming all CHO use is replaced by glycerol and ketones[a]

		Carbohydrate[b]		
Hibernation stage	Fat (g)	CHO budgeted (g)	Fat equiv. consumed (g)[c]	Protein (g)
Entry	4.6	16.0	10.0	5.8
Torpor	7.4	22.2	13.9	9.2
Arousal	8.5	29.6	18.5	10.6
Homeothermy	8.8	30.8	19.3	11.0
Total substrate consumed	29.3[d]	—	61.7[d]	36.6

a. The values summarized in this table are calculated on the basis of Table 5.1 and assumptions stated in the text (p. 88).

b. We assume that carbohydrate per se is not used at all but is replaced by ketones and glycerol, both of which derive from lipid. During entry, for example, the 16 g of carbohydrate are derived from 16/1.6 or 10 g of fat (see Table 5.3, note a).

c. 1.6 g carbohydrate is equivalent to 1 g lipid on the basis of the following calculation: 1 mol glucose (180 g/mol) yields 38 mol ATP (assuming the malate-aspartate shuttle) and 1 mol tripalmitate (860 g/mol) yields 24 mol acetyl CoA + 1 mol glycerol, which yields 287 mol ATP (each mole of ketone costs 1 mol ATP and 1 mol glycerol yields 11 mol ATP). 1 g glucose yields 0.21 mol ATP and 1 g tripalmitate yields 0.34 mol ATP; 1 g tripalmitate is equivalent to 1.6 g glucose in terms of ATP yield.

d. Total fat consumed is the sum of columns 2 and 4, or 91.0 g.

advantages of such a complex system (Figure 5.4) in the context of rodent hibernation. When amino acids are catabolized, dogma has it that NH_4^+ is the only potentially noxious end product, noxious because it can drive the glutamate dehydrogenase reaction toward glutamate and therefore deplete α-ketoglutarate in sensitive tissues like the brain. A certain amount of NH_4^+ can be tolerated and is recycled by way of glutamate dehydrogenase into glutamate; but if the nitrogen intake exceeds the demand, the NH_4^+ must be excreted. This is done via urea, which is synthesized by the urea cycle. In the hibernating rodent, there is no intake during dormancy, so the demand cannot be exceeded. So, why not simply recycle the NH_4^+ back to glutamate using glutamate dehydrogenase? Why involve the urea cycle, which requires four high-energy phosphate bonds per mole of urea, and the complicated cycling of urea into the blood, then the gut, and CO_2 and NH_4^+ back into the blood? The answer to this question can be found by looking more closely at the function of the urea cycle.

The urea cycle not only produces urea, a nontoxic form of NH_4^+, but in doing so prevents alkalosis of the tissues that would otherwise be the result of protein catabolism (Atkinson and Camien, 1982). The oxidation of any carboxylate anion, such as an amino acid, at physiological pH must produce H_2O, CO_2, NH_4^+, and HCO_3^-. Water is not a problem (Guppy and Ballantyne, 1981), and CO_2 can be excreted across the lungs. The nonvolatile wastes NH_4^+ and HCO_3^-, however, cannot be disposed of harmlessly; and if NH_4^+ were simply recycled in the animal, HCO_3^- would build up and pH would rise, with lethal effects. The urea cycle solves this problem by consuming 2 mol of HCO_3^- and 2 mol of NH_4^+ and producing 1 mol of urea, 1 mol of CO_2, and 3 mol of H_2O.

In a conventional mammal the urea and thus the offending HCO_3^- is excreted in the urine. Urine production in the torpid ground squirrel however, is minimal (down by 90–95%), and urine is certainly not excreted during torpor. A tempting role for the complicated urea cycling system is therefore the removal of HCO_3^- from central tissues to the gut. This system costs two ATP equivalents per mole of NH_4^+, but this is only 13% of the ATP that is produced by the oxidation of alanine, a small amino acid. This is surely a worthwhile cost when the benefits are the recycling of valuable nitrogen and the maintenance of pH stability. This assigned role must remain tentative, however, because of the lack of data concerning the fate of HCO_3^- in the gut. Shunting HCO_3^- to the gut is only a stop-gap measure, and net elimination is necessary in order for the proposed role of urea cycling to be a feasible one. The HCO_3^- in the gut can be eliminated from the body via two possible routes (see Figure 5.4). Bicarbonate is in equilibrium with water and carbon dioxide, as a result of both the uncatalyzed and the carbonic anhydrase-catalyzed dehydration–hydration cycle. The proton source for this reaction is a stumbling block, however. In fact, the production of the necessary protons is thought to be a significant role of the reactions of the urea cycle (Atkinson and Camien, 1982). The significance of carbon dioxide in expired air as an elimination pathway for HCO_3^- in the gut therefore is unknown, but probably small. The only other feasible route is through insoluble carbonates in the feces after hibernation. This route is theoretically possible, but carbonates have not been measured in the feces as the animals come out of hibernation.

Torpor has so far turned out to be surprisingly problem-free. A few qualitative changes in metabolism are evident, to cater for specialized needs, but the major anticipated problem, that of starvation, is simply absent, as a result of the dramatic decrease in metabolic rate. Fat and protein reserves are more than ample, and NH_4^+ accumulation and fuel for the brain are problems that can be solved easily in a number of ways. What then, are the factors that obviously become limiting during torpor and cause periodic arousal? Perhaps a related question is, What causes death from hypothermia in nonhibernating mammals?

Cold Death in Endotherms

Most mammals, as mentioned in Chapter 4, suffer from hypothermic cell death; hibernators obviously do not. The search for the basis of this resistance continues and has covered the areas of temperature and enzyme–substrate binding, ATP levels, and membranes.

Enzyme–substrate binding. Apart from the Q_{10} effect of temperature on enzyme activity, which is due simply to the effect of temperature on the movement of molecules, the bonds that determine the enzyme's function are also affected by temperature. Ionic bonds strengthen as temperature drops; hydrophobic bonds weaken. Enzymes employ both types of interactions both for enzyme–substrate complexing and for bonds integral to the protein structure itself. Thus the effect of temperature on enzyme structure and enzyme–substrate affinity depends upon the proportion of ionic to hydrophobic bonds present. It is known that an enzyme designed to work at one temperature becomes less efficient at another temperature as a result of the

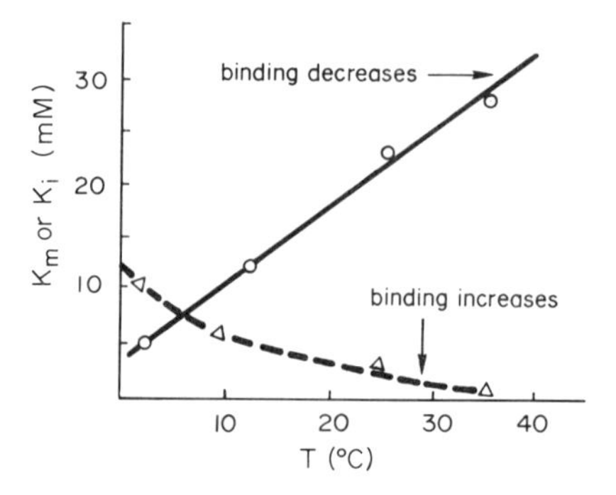

Figure 5.6 Effect of temperature on ionic and hydrophobic binding forces. The binding of the inhibitor dimethyl ammonium (○) is largely determined by electrostatic forces. The binding of the uncharged 3,3-dimethylbutyl acetate (△), a carbon analogue of the true substrate for acetylcholinesterase, is largely determined by hydrophobic forces. After Hochachka, 1973.

unsuitable proportion of ionic to hydrophobic bonds (Figure 5.6). These sorts of restraints could account for hypothermic cell death in nonhibernating mammals. There are no data on the contributions to binding in the enzymes of hibernators, so the role of adaptations at this level is unknown.

ATP levels. Many studies of the effects of temperature on primary metabolism have used ATP levels, and occasionally energy charge, as indicators of perturbations. Results consistently show that low temperatures have little effect on adenylate levels in both hibernators and nonhibernators (Lyman et al., 1982; Willis, 1979).

Membranes. The lipid composition of membranes changes with temperature, as do the activities of the appropriate lipogenic pathways. The two major consequences of these changes are perturbations of the activities of membrane-bound enzymes and of the membrane ion transport processes (Hazel, 1973). The former has not been characterized, but the latter, as mentioned in earlier chapters, is the favored locus for the imbalances that result in cold-death. The sodium pump is relatively temperature-insensitive in hibernators, and the reduction in passive fluxes of K^+ induced by low temperature is more marked in hibernators than in nonhibernators (Willis, 1979; Willis et al., 1980). Consequently, during hibernation, K^+ levels remain stable in the liver, diaphragm, ventricle, cerebral cortex, and erythrocyte of the rodent, ensuring continued function of these vital tissues. The thigh muscle, however, loses K^+ during hibernation, and from these data comes the only credible hypothesis concerning the triggering of periodic arousal (or the limiting factor during torpor) in rodent hibernators. The loss of K^+ from excitable cells would ultimately have the effect of depolarization of muscle and nerve cells. The depolarization would bring the membranes closer to threshold and therefore would lead to an increase in excitability (Willis et al., 1971). An increase in excitability, or irritability, as hibernation proceeds, has been reported by Lyman and O'Brien (1969). The trigger mechanism of periodic arousal therefore may be based on excitable tissues losing K^+. These are the obvious tissues to use, because, unlike most tissues, they will react to ion imbalances by disturbing the animal without endangering its survival, and they are robust and therefore can tolerate and fully recover from such metabolic imbalances. Presumably the function of periodic arousal is not to correct the ion imbalances of skeletal muscles but to metabolically forestall the same event in the more vital tissues. A related hypothesis has been proposed by Strumwasser et al. (1964), according to which neurons become gradually more sensitive to preexisting levels of neurotransmitters during hibernation.

So far, the rodent hibernator has managed with ease the feat of entering torpor and has, again with ease, survived the low temperatures and the starvation aspects of torpor. But entry into, and enduring torpor would be of no value if the animal were unable to return to the homeothermic state. The arousal process, in terms of kilocalories per hour, is the most demanding period of the hibernation cycle.

Initiation of Arousal

Arousal takes about 2.5 hr and is characterized by rising T_b, heart rates, and blood pressure and by extraordinary O_2 consumption rates, which can exceed 1000 ml/hr for a 250–300 g ground squirrel (3.5 ml $g^{-1}hr^{-1}$) (Lyman et al., 1982; Wang, 1978; Figure 5.7).

The question of initiation of arousal is inextricably linked to the factor that limits torpor, an aspect that has already been discussed. According to Twente and Twente (1978), the whole hibernation cycle is under autonomic control. Parasympathetic factors, which dominate during torpor, are gradually overcome by the sympathetic system late in the torpor period. Therefore the initiation of arousal, and the accompanying increased heart rate and vasoconstriction, are mediated through the sympathetic system. Such a mechanism could exist within the framework of the K^+ leakage hypothesis discussed earlier, resulting in the limits of torpor abutting neatly against the initiation of

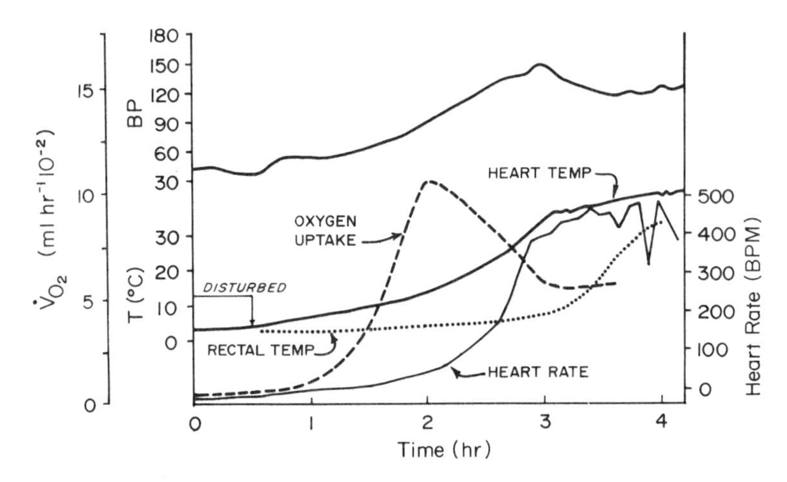

Figure 5.7 Blood pressure (in mm Hg), temperature, and heart rate of a thirteen-lined ground squirrel arousing from hibernation (Lyman, 1965). The O_2 uptake data is from a Richardson's ground squirrel (Wang, 1978).

arousal. The initiation of arousal is not of great metabolic significance, however, and will not be discussed further.

Metabolism of Arousal

One purpose of arousal is to heat the animal to 38°C. The animal then remains at this temperature for 8–10 hr, presumably correcting the imbalances that limit the torpor period.

Heating. Fat is certainly the major source of heat during arousal. RQ values remain low (0.76), fat deposits are depleted, plasma free fatty acid levels rise 4-fold, and glycerol rises transiently (Snapp and Heller, 1981; Spencer et al., 1966; Galster and Morrison, 1975). Brown fat deposits are present in the rodent hibernators and are definitely depleted during arousal. (For a review of brown adipose tissue function, see Cannon and Nedergaard, 1982.) Shivering is also evident, but the contribution by, and the necessity of, shivering and nonshivering thermogenesis is unknown (Lyman et al., 1982; Grodums et al., 1966; Dempster et al., 1966; Spencer et al., 1966; Joel, 1965). Using the data given in Table 5.1 (active ground squirrel column), and assuming a 3-hr arousal period at 800 ml of O_2 per hour per animal, we calculate that 1.2 g of fat would be utilized during arousal (this assumes that lipid was also the source of carbohydrate, through glycerol and ketones; see the footnotes of Table 5.2). The measured value of brown fat utilization during arousal is 0.69 g (Joel, 1965). Brown fat reserves in a 185-g ground squirrel are about 4.5 g (Joel, 1965), which *alone* could support an O_2 uptake of 3000 ml/hr for 3 hr. So brown fat reserves alone can easily support the heating phase of arousal.

Protein. Oxidation of protein increases as the ground squirrel arouses from torpor, but maximal activities reached are not higher than those evident during normothermia (Whitten and Klain, 1968; Galster and Morrison, 1975; Whitten et al., 1974). This increase therefore reflects only a quantitative (not a qualitative) change. Protein synthesis, and thus cell repair, appears to be delayed until the normothermic period when amino acid incorporation is at its peak (Whitten and Klain, 1968).

Carbohydrate. Gluconeogenesis and liver glycogen remain the running sore of the whole scheme. As mentioned in the section on carbohydrate and urea, some rodent hibernators deplete liver glycogen and some do not; as a result, data concerning gluconeogenesis during arousal are conflicting. It has been demonstrated that glycogen levels are partially restored during arousal (Galster and Morrison, 1975) and that glucose incorporation into liver glycogen is increased during arousal (Tashima et al., 1970). But, glycogen production from amino acids appears to drop during arousal (Whitten and Klain, 1968). So, if the worst case is taken, that liver glycogen (0.2 g) is completely depleted during torpor, is enough fat or protein or both mobilizable during the arousal and normothermic period to replace this glycogen? To replace 0.2 g of glycogen, either 2.0 g of fat (to supply the glycerol), or 0.35 g of protein would have to be oxidized (see Table 5.3). During a 3-hr arousal (800 ml O_2/hr) and a 10-hr normothermic period, 2.4 g of fat would be oxidized (assuming that ketones account for the carbohydrate component; see Tables 5.1 and 5.2). So glycogen levels could be restored using only glycerol, leaving protein free for repair or oxidation or both.

But is glycerol a feasible gluconeogenic precurser, and what of the NH_4^+ produced by the amino oxidation? Even though glycerol kinase is an uphill reaction in the direction of glycerol 3-phosphate, (1) glycerol is as good a gluconeogenic substrate as lactate in rat kidney cortex (Newsholme and Start, 1973, p. 280), (2) the $K_{m(\text{glycerol})}$ of gluconeogenesis in rat liver is 0.5 m*M*, and (3) plasma glycerol levels rise to 0.5 m*M* in the rodent hibernator during arousal (Newsholme and Start, 1973, p. 286; Galster and Morrison, 1975). Urea nitrogen levels do rise in the normothermic animal (Galster and Morrison,

1975), and nitrogen is actually lost, but how much nitrogen is recycled and where the carbon skeletons would come from are unknown.

Hibernation: Costs, Benefits, and Lessons Learned

Finally, now that the major divisions of hibernation have been identified, explored, and at least partially understood, a budget can be constructed for the whole fasting period. Assumptions will be that (1) the total hibernation period will be six months long, which consists of 22 × 20-hr entries, 22 × 7-day torpors, 22 × 3-hr arousals, and 22 × 10-hr homeothermic periods and (2) O_2 uptake (ml/hr per 300-g ground squirrel) for entry, torpor, arousal, and homeothermy, will be 65 (Wang, 1978), 9,800, and 250 (Tables 5.2, 5.3, and 5.4). The scheme proposed in Table 5.3 is untenable. No ground squirrel has, let alone can afford, 204 g of protein without a balancing protein intake (Galster and Morrison, 1976). The scheme in Table 5.4 only adds an element of danger (from added protein depletion compared with Table 5.2), with no advantage gained. The preferred scheme, Table 5.2, is conclusive in suggesting that ketones, which yield over 90% of the ATP that can be produced by 1 mol of triglyceride, must

Table 5.3 Calculated carbon budget for a 300-g ground squirrel for a 6-month hibernation period, assuming all CHO use is replaced by glycerol

		Carbohydrate[a]			
Hibernation stage	Fat (g)	CHO budgeted (g)	CHO derived from fat (g)	Protein equiv. consumed (g)	Protein[b] (g)
Entry	4.6	16.0	0.46	27.3	5.8
Torpor	7.4	22.2	0.74	37.6	9.2
Arousal	8.5	29.6	0.85	50.4	10.6
Homeothermy	8.8	30.8	0.88	52.5	11.0
Total substrate consumed	29.3	—	—	167.8[c]	36.6[c]

a. We assume that carbohydrate per se is not used at all but is replaced by glycerol only, that is, from triglycerides (10% by weight of fat used). The remainder of the carbohydrate is assumed to be produced from protein, where 1 g of protein is equivalent to 0.57 g of carbohydrate. For entry, for example, 0.46 g of carbohydrate is produced from the oxidation of 4.6 g of lipid. Then the remaining carbohydrate budget (16.0 − 0.5 = 15.5 g) must be provided by protein, that is, 27.3 g of protein.

b. Protein required for normal protein catabolism.

c. Total protein consumed is the sum of columns 5 and 6, or 204.4 g.

Table 5.4 Calculated carbon budget for a 300-g ground squirrel for a 6-month hibernation period, assuming all CHO use is replaced by glycerol and ketones and liver glycogen is totally depleted at each entry

		Carbohydrate[a]			
Hibernation stage	Fat (g)	Fat equiv. consumed (g)	CHO budgeted (g)	Protein equiv. consumed (g)	Protein (g)
Entry	4.6	7.3	11.6[b]	—	5.8
Torpor	7.4	13.9	22.1	—	9.2
Arousal	8.5	18.5	29.6	—	10.6
Homeothermy	8.8	19.3	30.8	6.1[c]	11.0
Total substrate consumed	29.3[d]	59.0[d]	0.2[a]	6.1[e]	36.6[e]

a. Ketones and glycerol replace carbohydrate, as in Table 5.2. In this case, however, liver glycogen (0.2 g) is depleted completely during each entry, and is restored completely from protein and glycerol during homeothermy. Thus the total amount of carbohydrate used is 0.2 g, that is, the amount used during the first entry. The rest comes indirectly from fat and protein.

b. During entry, 0.2 g of glycogen is used each time. For 22 entries, 4.4 g are used. So the amount of carbohydrate that needs to be derived from fat is 16.0 − 4.4 = 11.6 g, which represents 11.6/1.6 = 7.3 g of fat (see Table 5.2).

c. During homeothermy, 4.4 g of glycogen must be regenerated; 0.9 g comes from the glycerol released from the 8.8 g of fat oxidized. The rest (3.5 g) comes from protein and represents 3.5/0.57 = 6.1 g of protein.

d. Total fat consumed is the sum of columns 2 and 3, or 88.3 g.

e. Total protein consumed is the sum of columns 5 and 6, or 42.7 g.

play an essential role in the metabolism of hibernation. If the results of Table 5.2 are compared with the results of Galster and Morrison (1976) (Table 5.5), the fat utilization in Table 5.2 is low, but protein depletion is comparable. The percentages of the total cost of hibernation accounted for by entry, torpor, arousal, and homeothermy (calculated from Table 5.2), are 16, 24, 30, and 31 respectively, compared with the data of Wang (1978), in which the figures are 13, 17, 19, and 52.

The lessons to be learned from the ground squirrels are these.

1. Protein is an inefficient fuel and is easily depleted. Table 5.2 and the data of Galster and Morrison (1976) indicate a 30–40% depletion of body protein, which is at least 60% of mammalian body protein available for fuel purposes (Lehninger, 1975).

Table 5.5 Comparison of calculated and measured fuel utilization

Fuel	Total used during hibernation season (g)[a]	Reference
Fat	106	Table 5.2
Fat	178	Galster and Morrison (1976)
Protein	43	Table 5.2
Protein	31	Galster and Morrison (1976)

a. Results from Table 5.2 multiplied by 7/6 to render them comparable to the 7-month hibernation period of Galster and Morrison (1976).

2. Lipid, because it is so reduced and can be stored so efficiently, is almost an unlimited resource and could easily share the oxidative burden of protein. In fact, if the protein utilization in Table 5.2 is reduced to equal that in Galster and Morrison (1976), the total weight of lipid needed to replace this lost protein is only 2.5 g.
3. Quantitative adjustments are powerfully effective. If the torpid ground squirrel remained homeothermic and used triglyceride to supply its energy needs, the 200 g of stored lipid would be exhausted in three months. To take a more extreme case, we can extrapolate from the metabolic rate of a deer mouse at 1°C in a 4 m/sec wind (Chappel, 1984). At this metabolic rate (weight-corrected), a ground squirrel using lipid to supply its energy needs would use up its 200 g of stored fat in only 2–3 weeks.

The Black Bear

Twenty-five years ago studies on black bears disclosed a drop in O_2 uptake and a profound weight loss during dormancy (Hock, 1960). In the early 1970s, studies on black, grizzly, and polar bears showed that a significant bradychardia, but only a slight drop in body temperature, occurred during dormancy (Folk et al., 1972). More recently, the black bear *Ursus americanus* has been the subject of comprehensive metabolic studies, which have revealed in the dormant bear a biological system that is active, yet, for perhaps five to seven months, completely closed and self-sustaining.

Metabolism during Torpor

Black bears begin denning early in October, and the torpor period usually lasts until April. During this time, the bear does not eat, drink, urinate, or defecate. Cubs are born in January and weigh between 250 and 400 g at birth and between 4 and 5 kg at the time of emergence (Nelson, 1973; Nelson, 1980). During the dormant period, body temperature remains between 31° and 35°C and heart rates drop from 50–60 beats/min to 8–12 beats/min (Folk et al., 1972; Nelson et al., 1973; Watts et al., 1981). The metabolic rate of a mammal, at 37°C, can be predicted using Equation (1).

$$\text{kilocalories/hour} = 0.676\, M^{\mathrm{b}}, \qquad (1)$$

where M is body mass in kilograms and b is set at 0.75. The *actual* metabolic rate of the bear during dormancy can be calculated from O_2 uptake data or from weight loss figures, assuming (1) that weight loss is due to lipid oxidation only, and (2) that the calorific value of lipids is 9.3 kcal/g. Measured metabolic rate (O_2 uptake) is 32% of the predicted rate (Watts et al., 1981). Metabolic rates, calculated from weight loss, vary from 44 to 178% of predicted values (Watts et al., 1981; Nelson et al., 1973). The quantitative aspect of dormancy metabolism is still a matter for conjecture and will be discussed later. The qualitative strategy is obviously a successful one, features a finely tuned interplay between fat, protein, and carbohydrate metabolism, and becomes evident before (and thus in readiness for) the hibernation period (Nelson et al., 1984).

Fat. Respiratory quotient (RQ) values, which approach 0.60 during the dormant period, indicate that fat is the major, if not the only, source of fuel during this period (Nelson et al., 1973). The importance of fat as a fuel for the bear is heightened by the paucity of mammalian glycogen stores (see Rodent section), which can only support limited metabolism for a matter of days, and by the constant lean body weight of the bear during dormancy.

Protein. Protein metabolism in the dormant bear is active, but qualitatively different from that during activity. Protein turnover increases 3- to 5-fold (Lundberg et al., 1976); the flux of alanine carbon into glucose, protein, and lipid esters increases, and the catabolism of alanine decreases (Ahlquist et al., 1976). Thus reciprocal changes in protein turnover and amino acid degradation take place during dormancy; and concomitant with these changes, nitrogen production in the urine drops to 20% of predormancy levels. The increased protein turnover could be due in part to increased demand for lipolytic enzymes and would also be autocatalytic, as more proteolytic enzymes are needed for increased protein turnover. The decreased, but definite, protein catabolism consists of both partial and complete degradation of amino acids (Nelson, 1980).

Urea and nitrogen. Fecal droppings excreted after the dormant period contain practically no nitrogen (Nelson et al., 1973), and during dormancy urea is reabsorbed from the bladder as fast as it is excreted into it (Nelson et al., 1975). Thus *no* nitrogen is lost during dormancy, and lean body mass therefore remains constant. The urea is of course recycled, as in the ground squirrel. The data suggesting urea recycling in the bear are more conclusive *and* more comprehensive than those for the ground squirrel. The urea produced during dormancy, which accounts for 93% of the nitrogen excreted into the bladder (Nelson et al., 1973), passes across the bladder wall unchanged into the blood. Urea turnover rates in the blood increase by 2- to 10-fold during dormancy and $^{14}CO_2$ is always detected in expired air after the administration of labeled urea (Nelson et al., 1975). The scheme in the bear is thus that depicted in Figure 5.4, except that the bladder is definitely involved, which is an uncertain point in the rodent.

Carbohydrate. Interest in carbohydrate metabolism during dormancy in the bear centers on glycerol, produced during triglyceride hydrolysis, and on ketone bodies, produced as a result of β-oxidation (Figure 5.5). Glycerol carbon can be traced to several amino acids, including alanine, and glycerol conversion to glucose, protein, and lipid esters is enhanced during hibernation. Ahlquist et al. (1976) and Nelson (1980) conclude that glycerol from triglycerides is used (1) to supplement carbohydrate supplies presumably for the brain, and (2) to provide the carbon skeleton for the recycling of ammonia from urea into alanine (Figure 5.8). Ketones are excreted into the bladder during dormancy (Nelson et al., 1973), and ketone levels in blood rise about 8-fold, to 0.2 m*M*. These values are considered insignificant when compared with the 500-fold rise and resulting 5.0 m*M* levels that occur in starving humans (Nelson, 1980).

So, during the five-month winter sleep, body temperature remains high, heart rate drops, and body fat stores supply ATP for body (not

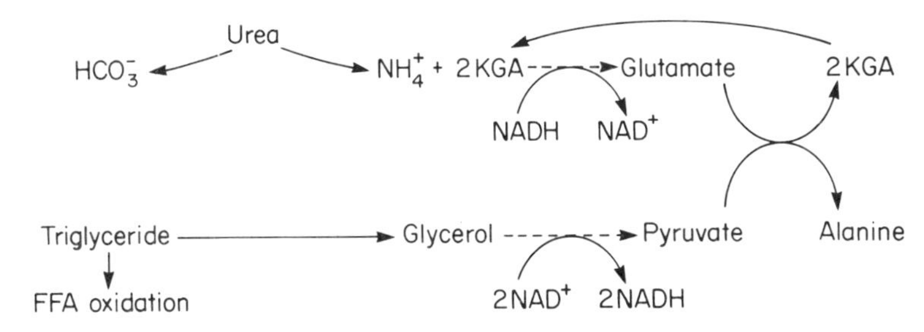

Figure 5.8 Nitrogen cycling in the bear. Relationships of reactions involving urea, glycerol, and alanine in the bear during torpor.

brain) maintenance metabolism. Glycerol and perhaps protein (which would explain the low RQ values of 0.60; Nelson et al., 1973) supply the needs of the brain, and any nitrogen produced is recycled back into alanine. The only input during hibernation is oxygen, the only outputs are heat and carbon dioxide and perhaps two cubs. This is the simplest interpretation of the data, and it turns out that it (1) underestimates the implications of the system and (2) reflects a misinterpretation of the data.

Urea cycling. Why is NH_4^+ cycled through urea in the bear during torpor? The answer is the same for the bear as it is for the ground squirrel—to prevent alkalosis. The principle is the same, but there is a quantitative difference between the bear and the ground squirrel. The bear, unlike the ground squirrel, does not periodically arouse and neither urinates nor stores urine during the hibernation cycle. Consequently, all the urea formed in the bear during torpor *must* be recycled. Because weight-specific O_2 uptake is at least as high in the torpid bear as in the torpid ground squirrel (see p. 97), unless protein catabolism is unusually low in the torpid bear, it follows that urea cycling activity must be higher in the bear.

Fuel for the brain

The problem of fuel for the brain during torpor occurs in the dormant bear as well as in the torpid ground squirrel. Weight-specific metabolic rates are similar in the two hibernators; and there is no reason to believe that the bear has unusually high glycogen levels in the liver. It is a characteristic of mammalian brain metabolism that O_2 uptake (of whole brain) is constant, whether the animal is sleeping or engaged in intense thought (Sokoloff, 1977). So stored glycogen could not conceivably supply the brain's fuel demands for more than a few days. What then fuels the brain of the dormant bear? In a starving man, glucose synthesized from glycerol and amino acids accounts for only 25% of the total fuel requirements of the brain. The remaining 75% is probably accounted for by ketones, whose total concentration in plasma rises by 500-fold to 5.0 m*M* after 8 days of starvation. In rat brain, the contribution by ketones to fuel requirements is less, 20% after 48 hr of starvation, by which time ketone levels in the blood have increased 15-fold to 3.5 m*M* (Newsholme and Start, 1973). Because lean body mass in the bear remains the same and no nitrogen is excreted, amino acids cannot be a *net* source of fuel, although amino acid carbon could end up as fuel for the brain. To produce glycerol from fat is a costly and inefficient business, in terms of glycerol produced per gram of fat. Even so, this avenue is worth further investigation. If the metabolism of the bear is supported totally by triglycerides containing 16-C saturated fatty acids, the brain, if fueled solely by the glycerol released through hydrolysis of these triglycerides, would maintain a metabolic rate that was about 4.5% of the total fat-fueled organism. Is this enough? It seem unlikely because (1) the brain of carnivores accounts for about 6% of the whole organism basal O_2 uptake, (2) the brain of the rat, which accounts for 5.7% of whole organism basal O_2 uptake, depends upon some ketone fuel during starvation, and (3) the glycerol would have to be shared between the brain and other tissues such as kidney and red blood cells, which also require a nonlipid source of fuel (Mink et al., 1981; Newsholme and Start, 1973; Robinson and Williamson, 1980). Levels of ketones in the blood of the active bear are 10-fold lower than in the fed rat, 0.02 m*M* compared with 0.2 mM; and they only rise to 0.2 m*M* in the bear during hibernation, compared with 3.5 m*M* in the rat after 48 hr of starvation (Nelson, 1980; Newsholme and Start, 1973). Even so, acetoacetate at 0.13 m*M* in the blood, is utilized by the brain of the fed rat (Hawkins et al., 1971), so there seems to be no evidence to suggest that the low levels of ketones in the blood of the dormant bear preclude their contributing to the fuel supply of the brain.

Low Ketone Levels: Costs and Benefits

How and why are ketones maintained at such low levels in the bear? In the rat, the rate of synthesis of ketone bodies is dependent upon the ratio of lipogenesis to lipolysis *and* upon the concentration of oxaloacetate. These two parameters appear to be controlled initially by the ratio of glucagon to insulin. Elevation of this ratio during starvation stimulates gluconeogenesis (which bleeds off oxaloacetate) and inhibits lipogenesis. The latter inhibition results in lower levels of malonyl CoA, an inhibitor of the enzyme catalyzing the regulated step of fatty acid oxidation. Starvation thus causes acetyl CoA production to increase and oxaloacetate availability to fall, changes resulting in increased shunting of acetyl CoA to acetoacetate (McGarry and Foster, 1980; Figures 5.9).

Although there are no significant changes in the concentrations of glucagon and insulin in the bear, nor a change in the glucagon:insulin ratio with the onset of hibernation (Nelson, 1978), the levels of ketones in the blood do rise 10-fold, a figure that is similar to the 15-fold rise in the starving rat. The important point is that ketone levels are abnormally low in the *fed* bear, 10-fold less than in the fed rat. A 10-fold increase in ketone levels during hibernation will *not* therefore pose a danger from ketotic acidosis, but will still provide the needed fuel for the brain. This feature of metabolism, rather than the cycling of nitrogen or the presence of huge fat reserves, probably represents the pinnacle of the bear's adaptive metabolic achievements. It is important to inquire how the ketone levels are kept low in the fed animal, and why levels do not rise higher during dormancy. Unfortunately, there are simply no data on the mechanism for keeping ketone levels low. Perhaps the activities of the enzymes producing ketones in bear liver are very low, or perhaps the production and availability of acetyl CoA and oxaloacetate respectively are extraordinarily tightly controlled, or both. Concerning the ketone levels during dormancy, there is considerable evidence suggesting increased lipogenesis in the dormant bear. Ahlquist (1976) found increased carbon flow from glycerol and alanine to lipid esters during dormancy, and triglyceride and cholesterol levels (free cholesterol is used in the polar skin of triglyceride micelles) rise in the blood of the dormant bear (Nelson et al., 1973). Increasing triglyceride synthesis and rising triglyceride and cholesterol levels are *not* hallmarks of starvation and could indicate a double-barreled mechanism in which (1) malonyl CoA levels are kept high to prevent excessive production of acetyl CoA, and (2) ketones are themselves substrates for lipogenesis (Robinson and Williamson, 1980), thus reducing their own level.

Even though the nature of ketone metabolism in the bear is unclear, the end result embodies the essence of succesful hibernation—a long-term, nonpolluting supply of substrate for the most sensitive organ in the body. But nothing is free, of course, so what are the costs

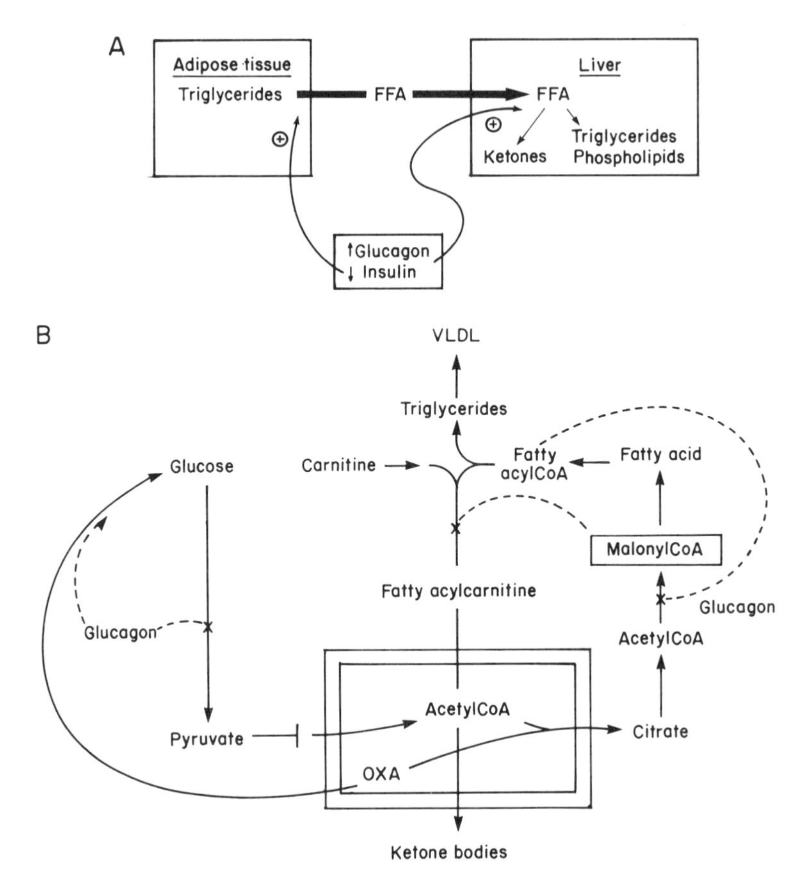

Figure 5.9 *A.* Bihormonal model for the control of ketogenesis. The model proposes that the mobilization of free fatty acids (FFA) from adipose tissue to liver results primarily from insulin deficiency. The activation of hepatic fatty acid oxidation and ketogenesis comes about through elevation of the glucagon:insulin ratio. After McGarry and Foster, 1980. *B.* Regulatory interactions between the pathways of fatty acid synthesis and oxidation in liver. In the fed state, malonyl CoA levels are high, assuring rapid fatty acid synthesis and suppression of fatty acid oxidation (through inhibition of carnitine acyltransferase I). Malonyl CoA concentrations may be lowered by glucagon excess or high tissue levels of fatty acyl CoA. In both cases the net result is cessation of lipogenesis and activation of fatty acid oxidation. OXA, oxaloacetate; VLDL, very low density lipoproteins.

of such a radical system? It is generally accepted that under "normal" conditions, ketone bodies are metabolic substrates in tissues such as adipose, brain, diaphragm, heart, kidney, and skeletal muscle. In fact, ketones are the preferred fuel of the heart and the kidney under certain conditions. Ketones are also used as a substrate for lipogenesis, mainly in adipose tissue, and play a regulatory role in utilization of glucose in peripheral tissues, production of glucose in the liver, proteolysis, and the supply and utilization of nonesterified fatty acids (Robinson and Williamson, 1980). Ketones could still play a regulatory role at lower concentrations, but ketone *utilization* seems to be largely regulated by their own concentration (Robinson and Williamson, 1980). This probably would not affect the brain during torpor, because ketone levels are 10-fold higher at this time and there is probably a carrier-mediated system, induced by starvation, that transports ketones across the blood-brain barrier (Conn and Steele, 1982). Nevertheless, a low contribution to fuel supply by ketones under normal conditions could well be the price of the bear's metabolic strategy during dormancy.

High Body Temperature during Torpor: Facultative or Obligatory?

The T_b of the bear during torpor is either actively maintained at 35°C, by juggling metabolic rate and heat loss, or it is simply an unavoidable consequence of the bear's size. A calculation based on spheres with known surface area:volume ratios and thermal conductivities suggests that the latter is the case (Figure 5.10). For these model calculations, the "ground squirrel" is kept at 5°C, in an ambient temperature of 0°C (a typical ambient temperature and T_b for a hibernating ground squirrel) by the generation of 4.89×10^{-4} joules $(\text{cc}\cdot\text{sec})^{-1}$. If the same amount of heat is generated, per cc, in the "bear," the "bear" is maintained at 33°C. Not only are these predicted temperatures close to measured temperatures, but O_2 uptake values (calculated using the value 4.89×10^{-4} joules generated $(\text{cc/sec})^{-1}$; and 1 liter of O_2 being equivalent to 4.7 kcal) are also very similar to measured uptake values during torpor (Table 5.6). Because the predicted (or measured) O_2 uptake of a torpid bear is only about 30% of that predicted by Equation 1 it seems (1) that metabolism is suppressed during torpor in the bear, but only by a factor of 3, compared with a factor of 30 in the ground squirrel, and (2) that despite the lowering of metabolic rate, because of the low surface area:volume ratio of the bear (Figure 5.10), T_b remains high. A pertinent question,

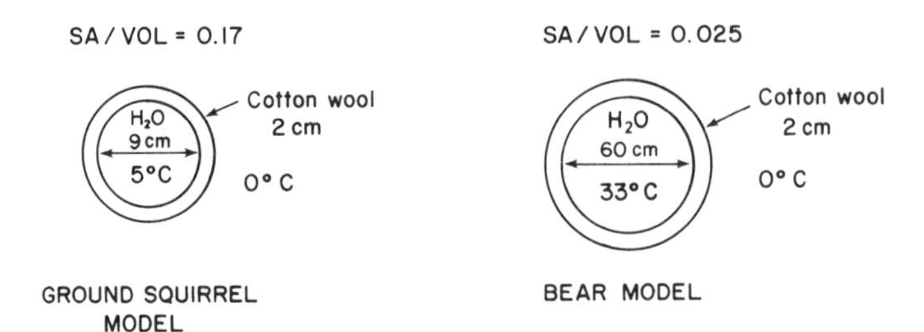

Figure 5.10 Effect of surface area:volume ratios on body temperatures of ground squirrels and bears during torpor. (1) The ground squirrel sphere corresponds to a sphere of water weighing 400 g (assumed weight of squirrel). (2) The bear sphere corresponds to a sphere of water weighing 100 kg. (3) Heat conductivity (K) of the cotton wool is 0.3×10^{-3} joules $(cm^2 \cdot sec \cdot °C)^{-1}$. Heat conductivity of fur is similar and varies between 0.2×10^{-3} and 0.6×10^{-3} joules $(cm^2 \cdot sec \cdot °C)^{-1}$ (Lentz and Hart, 1960; W.R. Phillips, personal communication). (4) Heat generated in the ground squirrel sphere $(S) = 3KT/tr = 4.89 \times 10^{-4}$ joules $sec^{-1} cm^{-3}$, where $T = 5°C$, $t = 2$ cm, and r = radius. (5) T in the bear sphere, using the S value in 4, equals $Str/3K = 32.6°C$.

but outside the scope of this chapter, is how does the bear keep its T_b at a mere 37°C during the active period of the year?

Advantages of Hibernation: Comparing Ground Squirrel and Black Bear Strategies

The ground squirrel and the bear represent two extremes among hibernators. The advantages of hibernation are clear in the case of the former; metabolic arrest enables the ground squirrel to survive an obligate fast, on fuel supplies (fat and protein) that would otherwise not be adequate. Is the situation as clear-cut in the bear? Bears are not really supposed to hibernate. Bartholomew (1981) suggests that a

Table 5.6 Predicted and measured O_2 uptake values in the bear and the ground squirrel during torpor

Animal	Predicted value (ml hr^{-1} $animal^{-1}$)	Measured value (ml hr^{-1} $animal^{-1}$)	Source of measured value
Ground squirrel	33	9	Table 5.1
Bear	8,250	5,600	Watts et al. (1981)

large animal cannot cool down because it would take too long to rewarm, but it is now clear that the cooling-down is the problem, not the heating-up. Swan (1981), quite convincingly, suggests that the only aspect of metabolism that can be turned down during torpor is that which is solely devoted to maintaining body temperature (M_{OH}). M_{OH} decreases with size and is absent at a body weight of about 110 kg. An animal weighing 110 kg cannot therefore turn down metabolism without affecting essential metabolism and thus cannot suppress metabolism during hibernation. This hypothesis appears to be supported by the data on the ground squirrel and the bear. The suppression of metabolism in the bear is an order of magnitude less than in the ground squirrel, yet both animal models turn metabolism down to the same level (presumably a minimal one), per g (Figure 5.10). But is the bear disadvantaged by its size? Up to a certain size we think not; in fact, we would argue that the large size is beneficial. The 60% metabolic suppression in the bear reduces weight losses from a prohibitive 45–90%, to 15–30% (Watts et al., 1981; Nelson et al., 1973), but *does not involve a significant drop in T_b and therefore does not necessitate homeothermic recovery periods*. But in very large bears, 110 kg and over, if Swan is correct, metabolic suppression is negligible, so what, in these cases, are the advantages of hibernation? Presumably fat stores would be large enough in these bears to support an unchanged metabolic rate for three to five months—but what about protein? In the ground squirrel, protein stores are limiting; and this problem is countered by urea cycling *and* metabolic suppression. In the large bears, however, metabolic suppression plays a negligible role in conserving protein stores, and thus urea cycling is the more important solution. Consequently, urea cycling is essentially 100% efficient in the bear, which is not the case in the ground squirrel. So hibernation in the bear (1) may result in some metabolic suppression with its concomitant energy saving, but (2), more important, it triggers the operation of the urea cycling pathway, which staves off near-certain death through protein wasting. The restraints of size again rear their head, as they so often do in biology, and force the bear to adopt a more restricted solution to the problem of winter fasting.

Summary

The nature of the mammalian hibernation strategy illustrates two biological principles: first, a recurrent one, that metabolic depression is an effective and ubiquitous way to slow biological time; second, that hypothermia is potentially lethal, even in the mammalian hibernators. Rodent hibernation involves a severe drop in both metabolic rate *and* T_b. Entry into torpor is facilitated by perturbing heat production and heat loss mechanisms and may rely in some part on tissue-specific, temperature–pH interactions. Metabolic suppression is therefore affected through high Q_{10} values in heart and liver—but through respiratory acidosis in muscle and blood. The torpid period is one full of obstacles. The problem of metabolic fuels is solved by a combination of laying down fat stores, suppressing metabolism, and producing ketones for the brain. Urea and bicarbonate, the products of amino acid degradation, are recycled and excreted, respectively. The problem of hypothermia-induced ion imbalance is only partially solved. The Q_{10} of the sodium pump and passive K^+ leak are decreased and increased respectively which does allow, in mammalian terms, a long survival time at remarkably low T_b. The effects of hypothermia are only delayed, however. As a result, each actual torpid period is limited to about 7 days, after which a homeothermic recovery period is needed. The black bear in many ways adopts a strategy similar to that used by the rodents. Fat stores are increased before hibernation, metabolism is depressed during torpor, and fatty acids and ketones supply the bulk of the fuel for primary metabolism during the hibernation period. Size, however, results in both obligatory and functionally significant differences between the hibernation metabolism of bears and rodents. Metabolic suppression in the bear is small compared to that in the rodent as a result of the lower nontorpor metabolism in the bear. Consequently, the T_b of the torpid bear is high and the torpid period is not limited by ion imbalances caused by hypothermia. Hibernation in the black bear does result in fuel conservation, but the major advantage appears to be the triggering of a 100% efficient urea-cycling pathway, which effectively curtails a prohibitive protein-nitrogen loss. In the case of the rodents, time extension converts 1 g of fat into 2–8 g of fat, but comes with the rider that the hypothermic periods must be limited in time. The bear does not really use metabolic suppression as the only mechanism of extending time. Rather, it uses a qualitative solution, which in terms of nitrogen loss, extends time, as far as we can tell, indefinitely.

6

Estivators

Tolerance to complete loss of extracellular and intracellular bulk water, as occurs in anhydrobiosis (see Chapter 9), represents an obviously extreme adaptive response to water lack, so we are not surprised to find that it is not a realistic strategy for most higher organisms. Nevertheless, many higher organisms are most adept at surviving seasonal periods of drought. Several broad patterns of adaptations have been noted by workers in this area. In one strategy, common in Arthropods, for example, life history is adjusted so that the species is represented only by eggs during the most severe part of the year. A second strategy, particularly common among desert mammals, is behavioral avoidance of the problem (seasonal migrations, living in underground burrows, adopting nocturnal activity cycles, and so forth). Neither of these bear upon the theme of our book. A third strategy does, however; and it is utilized by numerous invertebrate and vertebrate species routinely surviving stressful periods of water limitation. These organisms have developed mechanisms allowing them to tolerate certain degrees of dessication but to go dormant when dessication becomes too severe. Desert snails, for example, are able to survive seasons, or even years, of no rain by withdrawing into and closing off their shells with either unusually thick epiphragms or unusual numbers of them (Schmidt-Nielsen et al., 1971). The same strategy is utilized by Puerto Rican snails living in open tropical lowlands that undergo periodic but severe drought (Heatwole, 1983).

The insect world is particularly filled with examples of species that routinely enter estivation (or diapause) to survive periods of drought. Perhaps because of the agricultural importance of insects, estivation has been described widely across this class of organisms (Mansingh,

1971; Riddiford and Truman, 1978; Jungreis, 1978). From such studies it is known that entry into diapause can occur at any of the life stages of insects: embryonic and larval diapauses are induced by the *occurrence* of a specific hormone (a proteinaceous diapause hormone controls diapause in embryo stages and the juvenile hormone controls it in larval ones), whereas pupal and adult diapauses depend upon the *absence* of particular hormones (the prothoracicotropic hormone in the pupal stage, and the juvenile hormone in adult stages). Among vertebrates, the lungfishes of Africa and South America are particularly well known for being able to estivate during periods of drought (Hochachka and Randall, 1978), but numerous amphibians, reptiles, and small mammals are also able to thrive in dry, or even desert, environments.

In all forms of estivation a number of common features emerge, including (1) extremely low metabolic rates, (2) hypophagia and inactivity, and (3) programmed sets of biochemical and physiological adjustments for protecting the organism against lack of water and other harsh environmental parameters. The magnitude of these adjustments, by usual metabolic standards, can be immense. The dormant pupa of the *Cercropia* moth, for example, sustains one of the lowest metabolic rates of all animals, whereas an active adult (a closely tuned flying machine) displays one of the highest, with an *absolute* 2000-fold difference potentially separating these two extremes.

Whereas the magnitude and biological roles of estivation are well appreciated, underlying molecular and metabolic mechanisms largely remain a mystery. The best-studied estivating systems are insects, small mammals, lungfishes, and terrestrial toads. Because mechanisms of insect estivation are similar to those required during overwintering dormancy and the mechanisms of mammalian estivation are similar to those in hibernation, we will not deal with these topics further in this chapter; instead, we will focus on the lungfishes and toads.

Lungfish Estivation

Although in other lower vertebrates temperature and photoperiod may also be used as signals for entry into estivation, in lungfishes the only stimulus for entry into estivation appears to be lack of water. Usually lungfishes live in large bodies of water that are relatively stable on a year-by-year basis; in this environment, there is never a need to estivate. However, they often also occur in smaller lakes, ponds, and streams that may dry up during the dry season of the year. Even the larger lakes in which they live may seasonally vary in depth, changes converting what would otherwise be littoral zones to mud cakes. Thus, in various regions of their normal habitat, lungfishes may become trapped in pools or swamps that periodically dry up. Before such pools become hard, dry flats, however, lungfishes burrow into the mud, form a kind of cocoon with a breathing channel to the surface, and there enter an estivation period that can last for one dry season or for several years. Kjell Johansen has maintained an African lungfish in the estivating state for nine years, which is presumed to be a record.

Metabolic Rates and Organization in Lungfish

Lungfishes are not vigorous swimmers and seem specialized in their active phases of life as ambush predators. When they are active, they fuel metabolism with glycogen (glucose) or with fat; the first for anaerobic and aerobic metabolism, the second for standard oxidative metabolism. Glycogen is stored in all tissues, but as in all vertebrates, depots are highest in the liver. Fat, on the other hand, appears to be stored in two large masses running in an anterior–posterior direction and located primarily in the posterior third or so of the body (Dunn et al., 1983). In addition to fat and glycogen, normally active lungfishes routinely utilize proteins and amino acids as fuels for metabolism (Hochachka, 1980).

Enzyme and ultrastructural studies of tissues and organs of both the South American and African lungfishes are consistent with multiple-substrate-based metabolism. Whereas skeletal muscles have relatively high activity ratios of anaerobic to aerobic enzymes, the absolute activities for both groups of enzymes are substantially lower than for other, more robust fishes. Even the heart and brain of lungfishes display higher glycolytic potentials, but lower oxidative potentials, than those found in other fishes, whereas liver and kidney maintain a significant capacity for gluconeogenesis as well as for amino acid metabolism and the urea cycle (Hochachka, 1980).

The overall organization of metabolism, even in the nonestivating animal, therefore, is indicative of an animal "ticking" over slowly, which in fact is observed by direct measurements: $\dot{V}_{O_2}$ for a 500-kg lungfish at 22°C is estimated to be about 0.5 mmol hr^{-1}, compared with $\dot{V}_{O_2}$ of about 2.5 mmol hr^{-1} for a trout of the same size and at about the same temperature. During estivation, energy metabolism is even further reduced. Homer Smith in his original (1929) studies found the metabolic rate of estivating lungfish drops to only about one-third of normal. However, more recent measurements indicate that the metabolic rate during estivation continues to be depressed further and further with time of estivation and may drop by one to two orders of magnitude (K. Johansen, personal communication).

Glycogen Sparing

Simple calculations show that even at low metabolic rates, the amount of glycogen in lungfishes is inadequate to support long-term estivation. Moreover, as is observed in other stress situations as well (for example, salmon migration), glycogen and glucose reserves are maintained throughout estivation, presumably for those cells and tissues (brain, red blood cells, kidney tubules) that may have an absolute requirement for glucose; muscle glycogen may also be spared for use during emergency situations, as in arousal from estivation (for burst swimming). Because lungfishes in cocoons are clearly fasting, glycogen reserves must be maintained by gluconeogenesis from protein-derived amino acids. Although regulation of gluconeogenesis is not well worked out, the regulatory properties of at least one control site enzyme in the process (FBPase) are similar to those found for the enzyme in other fishes and mammals. Supporting this activity also are ample levels of the enzymes—glutamate-pyruvate transaminase (GPT), glutamate-oxaloacetate transaminase (GOT), and glutamate dehydrogenase (GDH)—required for the mobilization of amino acid carbon toward glucose (Hochachka, 1980; Dunn et al., 1983).

Fuel Preferences and Utilization Order

Homer Smith first showed that during the early phases of estivation, lungfishes fuel metabolism with a mixture of substrates and respiratory quotients (RQ values) are therefore near 0.8. To date, no one has estimated how long lungfishes can estivate using fat as a sole carbon and energy source, although it is probable that an African lungfish could easily estivate through a dry season mainly on fat stored in its lateral tail depots. Assuming 50 g of triglyceride in a 500-kg lungfish, we calculate that at a $\dot{V}_{O_2}$ of 0.5 mmol hr^{-1} the animal could estivate for about 240 days on this reserve alone. Because metabolic depression becomes more severe as estivation is prolonged, this time period in theory can even be longer, so fat stores are a potentially important fuel reserve. Moreover, there are advantages to fat as a fuel during estivation: (1) fats are highly efficient in terms of ATP yield per mole of starting substrate, and hence at the low metabolic rates required during estivation fats should be able to fire the animal's metabolism longer, on a molar basis, than any other substrate; and (2) fats yield no more noxious an end product than CO_2, which, of course, can be exhaled. Unlike the situation in some vertebrates, including man during starvation, in the fat-primed estivation of the lungfish there is no accumulation of ketone bodies. For these reasons we conclude that preferred carbon and energy sources during most bouts of natural estivation are fats. However, protein catabolism is needed for other reasons, and in the extreme the animal necessarily turns to endogenous protein as its main fuel source.

The Role of Protein and Amino Acid Metabolism

Despite the fact that protein and amino acids have been recognized as utilizable carbon and energy sources in the lungfish for 50 years, there are no available detailed studies concerning the mechanisms of mobilization of these metabolites. What is the major protein source that is mobilized? What tissues initiate the mobilization? What amino acids are utilized and where? The only working model available for fasting in fishes derives from studies of the spawning migration of salmon, and it supplies at least tentative answers to most of the preceding questions. The experimental basis for this model of metabolism in fasting fishes can be summarized as follows (Mommsen et al., 1980): white muscle, because of its large mass, is the primary storage house of utilizable protein required during fasting. It is the tissue sustaining the largest activation of proteolytic machinery and the largest depletion of endogenous protein reserves. As a function of fasting time, most metabolic enzymes show a continuous decrease in activity per unit wet weight of tissue. It is an instructive observation that cytosolic enzymes, such as LDH, and mitochondrial marker enzymes, such as citrate synthase, are utilized at similar rates throughout fasting.

The rate at which these metabolic enzymes decline in activity is similar to the rate at which soluble proteins in general decline in concentration, particularly during prolonged fasting; during early stages, metabolic enzymes appear to decline at slightly faster rates than soluble proteins taken as a whole. Nonsoluble proteins decline at the slowest rates in later stages of fasting. Thus, despite some selectivity in terms of rate, most proteins in white muscle seem to decrease in concentration presumably because they are all targeted as substrates for fasting metabolism.

Whereas a general loss of enzyme protein may be the rule, there are some exceptions; among metabolic enzymes, two such are now identified. These are glutamate-pyruvate transaminase and malic enzyme (ME). Among hydrolytic enzymes, the lysosomal group do not behave typically. Although white muscle proteolysis involves a selected activation of lysosomes, not all components respond the same way. Cathepsin D, for example, shows nearly a 10-fold increase in activity during fasting; at the same time carboxypeptidase A displays only a modest increase in activity; and β-*N*-acetylglucosaminidase shows no change at all.

From the preceding considerations we conclude that control of protein mobilization in fasting fish occurs at least at two levels (at the tissue level, where white muscle proteins are preferentially utilized; and at the lysosomal level, where only certain proteinases are activated). We assume that a tertiary level of contol is also operative; otherwise it would be difficult to understand how some enzymes are conserved while most metabolic enzymes are rapidly utilized—in fact at faster rates than soluble proteins all taken together, which are in turn hydrolyzed at somewhat faster rates than nonsoluble proteins. This behavior could arise because insoluble proteins are poorer substrates for proteases than are soluble proteins, as has been demonstrated for cathepsin D in mammalian studies. However, such an explanation does not seem adequate to explain the conservation of specific metabolic enzymes such as GPT and ME in white muscle. These two enzymes occur in both cytosolic and mitochondrial form; but in white muscle, which does not contain abundant mitochondria, the bulk of the GPT and ME activities must be located in the cytosol, where many other soluble proteins decline in activity during migration. Therefore, unless these enzymes are selectively protected against proteolysis, it is a fair assumption that their activities are conserved by adjustments in synthesis rates, as in mammalian liver during conditions of activated protein utilization. Why these specific metabolic enzymes should be conserved is clarified by studies of free amino acid profiles.

Fate of Amino Acids in Muscle of Fasting Fish

The end products of proteolysis are free amino acids, and from studies of acid hydrolysates the percentage composition of free amino acids being released into white muscle is known. For example, on a mole fraction basis arginine constitutes about 5% of such muscle hydrolysates; lysine, 9%; alanine, 10%; aspartate, 11%; and so forth. Because white muscle metabolism is largely anaerobic, it is reasonable to assume that when protein mobilization is initiated the large pulse of free amino acids entering the solute pool cannot be fully oxidized in situ. We might therefore expect a massive perturbation of the free amino acid pool at such time and an overall amino acid composition comparable to that of a general muscle hydrolysate. But this does not occur. Instead, the fractional composition of only a few amino acids is drastically altered. After prolonged fasting, when protein mobilization is strongly activated, white muscle arginine constitutes only 2% of the total pool, not the expected 5%; white muscle aspartate constitutes 4% of the pool, not the expected 11%; muscle serine, isoleucine, and leucine similarly occur at about half the fractional molar levels expected from protein hydrolysis. This means that the pool size of these amino acids is in each case depleted down to one-third to one-half of the expected levels.

On the other side of the coin, alanine constitutes 20–30% of the white muscle amino acid pool, not the expected 10%, and glycine constitutes 12%, not the expected 8% of the total pool. These two amino acids are clearly augmented; in the case of alanine, by up to 3-fold. Although not all amino acids behave in this manner (lysine and leucine in particular constitute about the expected 12% and 5% respectively of the white muscle free amino acid pool), these are crucial insights, for they imply that even if white muscle does not have the oxidative capacity for the complete metabolism of the large free amino acid pool being generated by proteolysis, it does have the metabolic machinery for numerous metabolic interconversions that lead to augmentation of some amino acid pools and depletion of others (Mommsen et al., 1980). Recognition of this fact is a key element in understanding how the mobilization of protein in fasting fish is organized.

According to our current understanding (Figure 6.1) of pathways of amino acid metabolism, most of the free amino acid pool can be converted to alanine. Some, including glycine and serine, are directly convertible to pyruvate for transamination to alanine. Others, such as

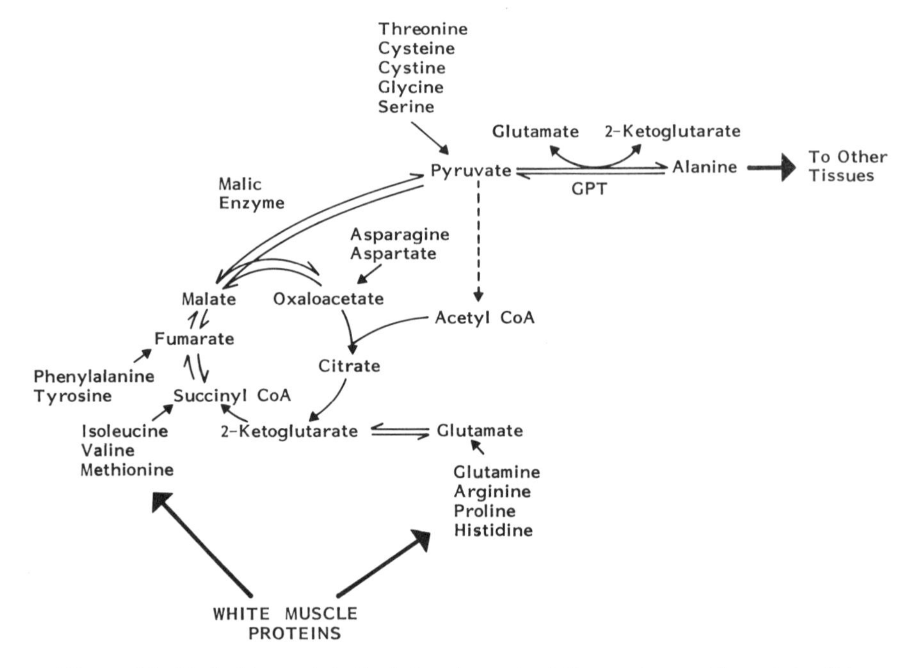

Figure 6.1 Probable metabolic fate of most of the amino acid pool in fish muscle during activated mobilization of white muscle proteins as major carbon and energy sources. From Hochachka and Somero (1984); reprinted by permission of Princeton University Press.

aspartate, can be converted to malate, then pyruvate, a process catalyzed by ME, thus again priming GPT-catalyzed transamination to alanine. Some, like valine, can enter the Krebs cycle through succinyl CoA, which is then converted to malate and to pyruvate and alanine. Still others, such as arginine, are metabolized via glutamate and a part of the Krebs cycle to malate, then to pyruvate and alanine (Figure 6.1). The overall quantitative energetics of these processes in white muscle are not known at this time, but would appear to be quite favorable. For example, the net conversion, glutamate → alanine, would yield 10 mol ATP per mole alanine formed, and this value may be even higher for proline and for arginine conversion to alanine. The favorable ATP yield from partial amino acid catabolism may indeed contribute to the muscle glycogen sparing that is observed in fasting fish.

Whereas ATP is a useful by-product of partial amino acid catabolism, it appears that the primary function of alanine is to serve as a

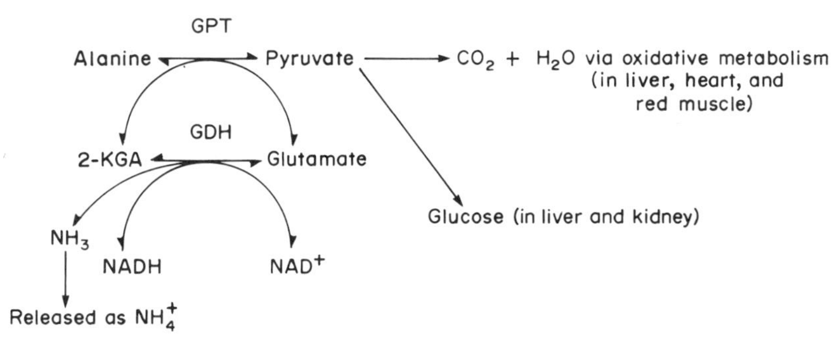

Scheme 6.1

carrier of amino acid carbon for further metabolism elsewhere, particularly in liver, red muscle, and heart (and probably in other tissues as well). From studies of nitrogen metabolism in teleosts, it is well established that the major pathway for the metabolism of alanine begins with GPT, whose function is linked to that of GDH as shown in Scheme 6.1.

Although this model explains why key enzymes in amino acid metabolism (GPT, and GDH in particular) are relatively elevated in tissues of fasting fish, it does not explain the ultimate fate of alanine-derived pyruvate. In many tissues, but most certainly in red muscle and heart, the predominant fate of pyruvate formed this way during migration is almost certainly oxidation. Indeed, in the extreme, the available data imply that alanine and other amino acids may become a major (if not *the* major) carbon and energy sources for fasting fishes.

In gluconeogenic tissues, such as liver and kidney, however, an important function of alanine is to serve as a glucose precursor. In fasting fishes (migrating salmon and estivating lungfishes), the liver sustains an enhanced gluconeogenic capacity (as judged by the activity of gluconeogenic enzymes and by direct flux measuresments); and it is well established that blood glucose supplies (for the brain) and glycogen depots (for white muscle work required at the end of estivation) are maintained at near normal levels. Endogenous supplies could not possibly explain this observation, so it is necessary to assume that glucose homeostasis is maintained by gluconeogenesis. Because of its availability, alanine, which is in equilibrium with pyruvate and lactate pools, is considered the most likely gluconeogenic precursor, a proposal consistent with the capacity of these three-carbon compounds to prime glucose synthesis in various fish liver preparations (Mommsen et al., 1980).

This model of protein catabolism in fasting fishes (Figure 6.1) not only accounts for all currently available data in this area, but also is useful in explaining data that would otherwise be perplexing. In the first place, it provides a rational metabolic explanation for controlled release of amino acids from white muscle, and it may explain why the free amino acid pools in white muscle and blood are not greatly perturbed during extensive proteolysis. The model explains why GPT and ME are maintained during fasting and why alanine has often been identified as a carrier of amino acid nitrogen and carbon from fish and mammalian skeletal muscle. For white muscle, the model supplies a critical function for ME under conditions of protein catabolism and may explain why the enzyme in some fish white muscle occurs at over 10-fold higher levels than in mammalian skeletal muscle. Because for liver, red muscle, and heart metabolism alanine is supplied from exogenous sources, the model also explains why key enzymes in amino acid catabolism (GPT and GDH, in particular) are elevated in these tissues in fasting fish without the concomitant activation of proteolytic enzymes to ensure an endogenous supply of amino acid substrate. Finally, for lungfishes, this model explains why key enzymes (GPT, GOT, GDH) required for this kind of metabolic process to work occur in reasonably high activities in all the appropriate tissues and indeed are elevated during estivation even while overall oxidative metabolism is potently depressed.

For the preceding reasons, we tentatively conclude that in estivating lungfishes (1) the main storage house of mobilizable protein is white muscle; (2) the mixture of amino acids released into white muscle by proteolysis does not merely spill out into the blood nor is it metabolized completely in situ (instead, a partial metabolism of amino acids leads to an enrichment of white muscle supplies of alanine, which, perhaps along with only a few other amino acids, form the true substrates for catabolism at other tissues and organs in the body); and (3) the main fate of alanine in some tissues is oxidation, but in the liver (and possibly the kidney) it is gluconeogenesis.

Nitrogenous End Products

Although details of our interpretation of amino acid and protein metabolism in lungfishes remain to be filled in, the data are unequivocal in regard to the end products formed. These are CO_2, H_2O, and glucose, on the one hand; ammonia and urea, on the other. The first three need no further discussion at this point, but the waste nitrogen problem is of crucial importance for one fundamental reason: like *Artemia* cysts, the estivating lungfish is largely a closed system with only O_2, CO_2, water, and heat being exchanged with the outside. Self-pollution is therefore an ever-present hazard, which is why the overaccumulation of potentially noxious end products (such as ammonia and urea) must be minimized.

The metabolic paradox faced by estivating lungfishes (of needing to mobilize proteins and amino acids and of needing to minimize the accumulation of their metabolic end products) was recognized by Homer Smith in his studies a half-century ago, and he obtained data that supplied at least an initial insight into how estivating lungfishes deal with it. His data, confirmed by later studies, indicate that lungfishes even in the active, nonestivating phase generate *both* ammonia and urea as end products of amino acid metabolism. On entry into estivation, the rates of ammonia formation decline but normal rates of urea production seem to be sustained. As a result, the ratio of urea produced to ammonia produced rises in estivating lungfish. Because urea is less toxic, most workers assume that it continues to accumulate through the estivation period in lungfishes. That may occur over the short term, but it is not very probable in the long term for a very simple reason: because the continuous accumulation of urea at the rates observed would yield blood urea levels of about 0.3 *M* in one year, 1 *M* in three years, and over 3 *M* in nine years. Such high concentrations are implausible, so it seems that something is still missing in our interpretation of this end products problem. The most probable missing concept in all previous discussions of the problem is that of urea recycling. However, whether or not urea recycling is utilized by estivating lungfishes, for the moment at least, must remain an unanswered question.

Overview

In response to water lack, the African and South American lungfishes harness the same survival mechanisms. Soon after a cocoon is prepared, to reduce evaporative water loss the animal enters a dormant period, which is maintained until the next rainy season arrives. Estivating metabolism is fueled by fat and protein catabolism, the latter being required to generate amino acid precursors for gluconeogenesis; tissues such as the brain presumably maintain an absolute dependence upon glucose as a substrate for energy metabolism. Maintaining a partially metabolically arrested state is probably facilitated by a strong dependence by the organism upon anaerobic metabolism even in the normal active state. This dependence arises necessarily from the low absolute activity levels of enzymes in energy metabolism and from unusually high anaerobic to aerobic metabolic potentials for most tissues and organs tested, including the brain. Also, because the organism is inactive during estivation, its single largest tissue mass (white muscle) is metabolically largely inert. In fact, the main metabolism of white muscle

proteins → amino acids → interconversion intermediates
→ alanine

probably adds a little to the estivating rates of ATP synthesis and turnover. Close inspection of the metabolic pathways involved indicates that the net reaction actually generates significant amounts of ATP per alanine formed, and this ATP generation may reduce dependence on carbohydrate and thus help to spare endogenous glycogen stores until estivation is terminated.

In lungfishes, as elsewhere, the main strategy is to enter a metabolically arrested state for the duration of the stress period. The degree of metabolic arrest is modest, however, for $\dot{V}_{O_2}$ is reduced to only about one-fifth of SMR during estivation bouts equivalent to about one natural dry season. A mole of substrate during estivation therefore can keep lungfishes "ticking over" five times longer. Or, put another way, for a bout of estivation equal to one dry season, biological time in the African lungfish is slowed down by at least a factor of 5.

Water Loss in Amphibians

The risks and hazards of dehydration are perhaps nowhere more acute than in amphibians living in dry areas. Because amphibians are ureotelic (excrete urea), they lack the renal water conserving mechanisms of uricotelic birds and reptiles (see below for two fascinating exceptions). They cannot produce hypertonic urine because the mammalian renal capacity for water conservation is unavailable to them. Furthermore, with a few exceptions amphibians sustain very high rates of evaporative water loss through the skin; evaporation is not subject to any effective control and the skin in most species exerts little or no resistance to evaporation. Consequently, rates of evaporative loss depend only on physical parameters (temperature, saturation deficit, and speed of movement of the air over the animal) and are influenced by the animal's activity, metabolism, and surface area.

In the absence of inherent and effective cutaneous control over evaporation, adaptations to dry habitats have developed along several pathways, including (1) behavioral adjustments; (2) increased rates of water uptake through the skin when water is available (amphibians never drink); (3) increased bladder volume, with dilute urine being stored and subsequently reabsorbed in time of water scarcity; (4) enhanced ability to survive loss of body water; (5) enhanced waterproofing; and, in the case of a few desert toads, (6) formation of an underground cocoon that retards water loss when the animal enters a dormant period of estivation. Of these potential ways of dealing with dessication, the last three are of particular interest to the theme of this book.

Adjustments in Water Loss Tolerance

Workers in this field recognize that, for an organism facing dehydration, even a low rate of loss may be significant if that species can tolerate loss of only a small proportion of its body water. The same rate of loss would be less significant for an animal with greater tolerances. To quantify an organism's tolerance to dehydration, the concept of the vital limit of water loss (VL) is used as a measure of desiccation tolerance. It is expressed as the percentage of the original fully hydrated weight that can be lost as water before death occurs.

Adjustment of the vital limit is one of the modes of dehydration adaptation in amphibians and reptiles (Heatwole, 1983). Values of VL between 25 and 30% are common, although in some desert species VL is over 50% of body weight. (In mammals, 10–12% is usually the lethal limit, although the camel can tolerate losing nearly twice this proportionate amount of water.)

From studies of dehydration–hydration cycles of isolated cells, it is evident that at a cellular level water loss per se need not represent insurmountable problems; after all, a fibroblast L cell is hardly desert-adapted, yet it can sustain greater than a 50% loss of intracellular bulk water (Chapter 9). The reader may therefore well wonder why animals such as amphibians and reptiles can sustain, on average, only a 25–30% desiccation. The reasons are probably to be found at integrated physiological levels rather than at cellular levels; for example, such severe dehydration would presumably lead to massive increases in blood viscosity, increases representing proportional increases in the metabolic costs of the cardiac pump. The mechanisms used by desert amphibians and reptiles to deal with these secondary problems of desiccation unfortunately are not known. In any event, it is clear that tolerating water loss even to the drastic degree of 50% of body weight is clearly a first line of defense. A far more effective strategy would be to minimize water loss or reduce the need for it altogether. Two mechanisms have been harnessed to achieve this end: the first is waterproofing, the second is burrowing and estivation.

Waterproofing Amphibian Skin Coupled with Uricotelic Habit

Waterproofing of the skin is not a strategy available to all amphibians, but it has been exquisitly developed in two exceptional genera. These two genera, *Chiromantis* (from Africa) and *Phyllomedusa* (from South America), are ecologically equivalent but phylogenetically divergent. Perhaps because they are in different families, the mechanisms for reducing evaporative loss are different in the two groups (Heatwole, 1983). In *Phyllomedusa,* waterproofing lipids are secreted by integumental alveolar glands, and these frogs wipe the secretions over their body with their feet. In *Chiromantis,* the lipid glands and wiping behavior are absent, and waterproofing seems to be related in some way to the presence of specialized chromatophore units. As a result of waterproofing, water evaporative rates are drastically reduced (to levels comparable to desert lizards, for example).

Waterproofing the skin would not be very fruitful if the frog then proceeded to lose copious quantities of water by other routes. As most amphibians excrete nitrogen waste in the form of urea, the loss of water that is required just to remove metabolically formed urea would be substantial. Whereas the advantages of minimizing water loss by this route would clearly serve as a driving force selecting for a water-conserving alternative metabolic pathway, the first report that *Chiromantis* excretes mainly uric acid—rather than urea, as in most adult amphibians—was somewhat of a sensation because it challenged long-held views of nitrogen excretory metabolism in the Amphibia. Nevertheless, in *Chiromantis* (Loveridge, 1970) 60–75% of the dry weight of urine is uric acid, whereas in *Phyllomedusa* (Shoemaker and McClanahan, 1975) fully 80% of total waste nitrogen is excreted as urate. Because urate is only slightly soluble in water (solubility is about 6 $\mu g\ ml^{-1}$), it can be excreted with minimal loss of water; that is why, the excretion of urate in insects, land snails, most reptiles, and birds has been traditionally viewed as an adaptation to water conservation in water-poor habitats. By excreting urate as its nitrogen waste, *Phyllomedusa* loses as little as 4 ml kg^{-1} day^{-1} compared with 60 ml^{-1} kg^{-1} day^{-1} that would be required for removing the same amount of nitrogen as urea—a not inconsiderable, 15-fold advantage.

Within each of these two exceptional genera, at least several species of frogs are known (Heatwole, 1983). Thus far all those investigated display low evaporative rates and excrete urate. Both of these genera are tropical to subtropical and both are arboreal. *Chiromantis xerampelina* inhabits low veld savannas from Zululand to Kenya, and *C. petersi* occupies arid, wooded steppe and subdesert steppe from lowland Ethiopia to north-central Tanzania. In South America, four species of *Phyllomedusa* occur in habitats ranging from open forest to xerophytic open forest and brush. Therefore, the habitat in which uricotelic, waterproofed frogs thrive is tropical to subtropical xeric, open forest or scrub; interestingly, only arboreal species in these habitats have developed such a unique water balance system. It appears that the combination of a relatively hot, dry environment accompanied by occupancy of a relatively exposed diurnal microhabitat (limbs of shrubs and trees in open forest or savanna) are the common ingredients in the coupled evolution of amphibian uricotelism and waterproofing in these remarkable frogs. Although an eminently successful dual mechanism, both components are ultimately limitable. On the one hand, urate cannot be excreted in a fully dry form, so there is a natural maximum limit to how much water can be saved by this mechanism. The maximum effectiveness of waterproofing, on the other hand, is set by a compromise that must be struck between water and thermal regulation.

Thermal Limits to Waterproofing in Frogs

At least in *Phyllomedusa* the advantages of waterproofing break down at high ambient temperatures, and there is good reason for this (McLanahan et al., 1978). Up to approximately 35°C these heat-tolerant frogs are thermal conformers, and the wax layer on their skin is highly impermeable to water because the wax is in a gel-like state. Between 34° and 38°C, however the wax begins to melt, and this phase transition marks the trade-off point between the benefits of water retention versus body temperature control. As the wax liquifies it retards water loss less effectively, so the frog "sweats," and this evaporative cooling allows the frog to hold its body temperature at

36°–37°C even if external temperatures go up to 41°C, a temperature close to the upper lethal limit for this species.

This arrangement is a useful short-term trade-off (increased water loss in order to maintain a body temperature low enough to be compatible with survival), but it too has its own limit. A frog is a very finite reservoir and cannot dissipate water (and thus heat) indefinitely. It is not surprising, therefore, that if these conditions are prolonged, the animal metabolically arrests and goes into a deep torpor. This final line of defense against intolerable dehydration levels is shared with numerous desert amphibians; because the process is somewhat better understood in the latter group, we will utilize them to illustrate the issue.

Estivation in Desert Toads

Spadefoot toads are among the most successful of desert amphibians, successfully surviving in deserts that are among the driest and hottest in the world. In Arizona the season of activity for three such toad species is initiated by erratic rains that usually arrive in July. After breeding, the toads feed for two to three months, then begin a nine- to ten-month period of continuous dormancy underground, where diurnal thermal extremes as well as the dehydrating effects of the surface are avoided.

As in lungfishes, an important feature (possibly a hallmark) of amphibians that routinely enter dormancy is that they are metabolically rather sluggish in the first place. In the case of the spadefoot toad, for example, the normoxic $\dot{V}_{O_2}$ at 15°C (about 20–30 μl g^{-1} hr^{-1}) is only one-fifth to one-fourth the $\dot{V}_{O_2}$ for frogs that do not estivate; this SMR displays a Q_{10} (about 2.3) that is in no way unusual (Seymour, 1973). During the feeding and reproductive period, the animal stores up fuels for later metabolism in dormancy. Although detailed studies of fuel utilization sequences are not available for dormant amphibians, it is known that the spadefoot toad uses fat during estivation, which is gradually depleted from initial levels of about 5% of body weight to about 2% after a nine-month bout of estivation.

How important metabolic arrest is to the estivating spadefoot can be illustrated with a set of simple calculations. At the low initial metabolic rate of this toad, lipid stores by themselves (equivalent to about 200 mmol palmitate per kilogram body weight) could sustain the organism for only about 153 days (at an O_2 uptake rate of 1.25

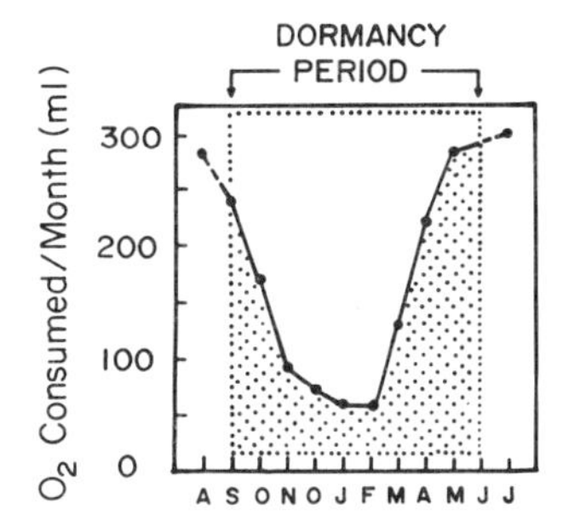

Figure 6.2 Total oxygen consumption of an hypothetical dormant *Scaphiopus* weighing 25.5 g, under field conditions. Modified from Seymour (1973).

mmol kg^{-1} hr^{-1}) or for only about 50% of the observed dormancy period. According to Seymour (1973), the metabolic rate of the spadefoot drops gradually on entrance into estivation, reaching after about three weeks stable values that average 0.25 mmol O_2 kg^{-1} hr^{-1}; that is, only some 20% of the normoxic, active rates (Figure 6.2). At this metabolic rate, the animal's fat stores alone could sustain it for about 765 days, or the equivalent of over two dry seasons. Thus the spadefoot retains a builtin energetic safety factor of at least 2-fold, with respect to this one substrate alone. Thus it is not surprising to find that after the equivalent of a single, nine-month dormancy period, only about half the fat stores are used up. For reasons mentioned in our discussion of lungfish estivation, it is almost certain that estivating toads must mobilize at least some proteins (for supplying amino acids as gluconeogenic precursors to allow glucose homeostasis). Thus the actual demands on fat as a fuel at this time may be even less than estimated by our calculations.

Possible Mechanisms of Metabolic Arrest

Although the usefulness of metabolic arrest is easily demonstrable for desert toads, mechanisms for switching down metabolism are not well understood. One possibility is that, as a result of hypoventilation and bradycardia, many and perhaps all tissues are at best mildly hypoperfused and that O_2 delivery is therefore also reduced. Because the $\dot{V}_{O_2}$ of tissues such as skeletal muscle is directly proportional to O_2 availability, this kind of pattern could possibly account for the metabolic depression observed in estivation. In this event, we would define these organisms as O_2 conformers with little or no capacity for activating the Pasteur effect (Chapter 2). Unfortunately, data are not available to help us decide whether or not these processes are important.

A second possibility is that, as in hibernating mammals, estivating toads may sustain better tissue-specific or global acidification, which in turn may serve as a generalized mechanism for depressing metabolism (Malan, 1978). There are good reasons for favoring this hypothesis. When desert toads burrow underground, they form soil-filled estivation cocoons around themselves to reduce evaporative water loss across the skin, but this simultaneously means a loss of the skin as a site of respiratory gas exchange. As a result, CO_2 tensions gradually rise and plasma pH gradually falls, presumably along with pH_i in most (or all) tissues. Whereas we might expect that increased gas exchange at the lung could remedy this imbalance and make up the energetic shortfall, estivating toads in contrast hypoventilate with coupled bradycardia. Thus the stage is set for a H^+-mediated metabolic arrest during the first few days of estivation. The preceding cascade of events, however, apparently does not continue indefinitely. At least in *Bufo marinus,* which appears to enter a more shallow dormancy than that found in desert toads, blood pH is regulated and the respiratory acidosis is progressively compensated for—probably by increasing mobilization of calcium carbonate reserves—so that within three or four days of estivation, plasma pH is returned to fully normal levels (Boutilier et al., 1979; Simkiss, 1968). Thus, even though the hypothesis of H^+-mediated metabolic arrest is attractive (because it implies a common and universal off-switch spanning organisms from *Artemia* embryo cysts to hibernating mammals), further work is needed on desert toads or on one of the arboreal species of *Phyllomedusa* to assess whether low and sustained pH_i in deep dormant species contributes to low rates of oxidative metabolism.

A third possible arrest mechanism, first suggested for hibernating mammals by Margules (1979), is that opiates are involved in the regulation of dormant states in vertebrates. The best and, to our knowledge, the only evidence in mammals in favor of this speculation is that injections of naloxone (an opiate-blocking agent) arouse hibernating hamsters. Analogous experiments with estivating *Bufo* by Randall and his coworkers suggest that a similar process may be at work in amphibian dormancy. Thus, intravascular injections of naloxone in saline into buried toads causes emergence from the burrow; the animal becomes hyperactive, concomitantly with an increase in both O_2 uptake and CO_2 production (D. J. Randall, personal communication). Implied in both studies of the dormant toad and hibernating hamster is a role for the opiates in dormancy. In terms of metabolic control, this is a very attractive hypothesis, because it helps to explain two other observations: (1) opiates are known to facilitate the release of glucagon and adrenocorticotropic hormone, both of which favor gluconeogenesis, and (2) endogenous opiates are also known to stimulate the release of antidiuretic hormone and thus play an important role in water conservation. As we have seen, both these functions are critical to estivating amphibians.

Summary

Amphibians living in deserts or in trees in hot, dry environments routinely are exposed to the hazards of desiccation. Several biochemical adaptations are used to avoid or minimize the problems of dehydration. First, even when active, these amphibians sustain relatively low metabolic rates, which automatically maintain relatively low evaporative rates. Second, many of these species expand their VL (vital limit of water loss) and are therefore able to tolerate a greater percentage of water loss. A third strategy, apparently restricted to South American and African aboreal frogs, involves the dual mechanisms of lipid-based waterproofing of the skin and uricotelism. The final line of defense, here as in the lungfish, is to enter a dormant, metabolically arrested state. Desert and South American aboreal amphibians are capable of deep dormancy, whereas some species (such as *B. marinus*) typically enter more shallow dormancy. In the former, oxidative metabolic rates are switched down to about one-fifth of normal. Protons may have a transient inhibiting role during early estivation but do not appear to have a pervasive role, because of stringent pH regulation during later phases of the dormancy.

For amphibians capable of dormancy, time seems to pass slowly even in the nondormant state; $\dot{V}_{O_2}$ of the desert spadefoot, for example, may be only one-fifth that of *Rana pipiens* at the same temperature. On entrance into deep dormancy, time is slowed down by yet another factor of 5—a metabolic arrest factor similar to that in the estivating lungfish. Thus 250 days of fasting in a dormant spadefoot is equivalent to about 10 days of fasting in *Rana pipiens* normally active at about the same temperature.

7

Frozen Insects

Temperate or polar terrestrial ectotherms seasonally or perpetually encounter temperatures potentially capable of freezing the ECF and ICF. Although the problem is common to many organisms, those organisms best understood are insects and some vertebrates. As we will see, the means for dealing with this problem differ somewhat in the two groups. Terrestrial insects have two general strategies for surviving freezing conditions: either (1) these organisms harness mechanisms to avoid freezing when exposed to temperatures below the freezing points of their internal fluids, or (2) they freeze and live on in the frozen state.

Although we will focus upon the latter strategy in this chapter, it is useful to briefly consider the former one as well. This strategy is utilized by species that are sensitive to freezing and they prevent the formation of internal ice by several well-understood mechanisms. These include: (1) depression of the supercooling point of body fluids, in some arctic species to −50°C, (2) elimination of ice nucleation sites in the body (for example, by emptying the gut), (3) accumulation of high concentrations of polyhydric alcohols as cryoprotectants, and (4) maintenance of antifreeze proteins in the hemolymph (Duman and Horwath, 1983). Although the majority of cold-hardy insects may overwinter in this manner, freezing avoidance based upon extensive supercooling is risky: spontaneous intracellular ice formation is a constant, potentially lethal hazard. Indeed, intracellular ice formation is presumably the selective pressure that drove some insect groups to develop freezing tolerance.

Freeze tolerance occurs among various insects (for example, beetles, flies, and wasps), with larvae or adults or both being able to freeze. Freeze tolerance is also common among plants, some helminths, and intertidal mollusks. Storey, Baust, and others (1981, 1983, 1985) have addressed this problem in the larvae of the gall fly, *Eurosta*, and much of our analysis is based on their recent studies. In this species, the third instar larvae overwinter inside stem galls on goldenrod plants; exposed above the snowline, the larvae face and survive the rigors of harsh northern winters, with temperatures sometimes falling to −50°C.

As in other freeze-tolerant species, oxidative metabolism is sustained at subzero temperatures as long as freezing does not occur. In *Eurosta*, adenylate concentration ratios remain normal and mainstream metabolic pathways (for example, proline and polyol synthesis) remain functional to −8°C. The balance shifts when freezing occurs, because the water→ice phase transition severely depresses metabolism: oxygen comsumption drops to less than 1% of normothermic rates, polyol synthesis stops, and adenylate energy charge drops. Glycolytic energy production continues at a low level, however, and lactate production in the frozen state is found in *Eurosta* and in other species as well. As in preparatory phases of anhydrobiosis, the metabolic processes functional during entrance into the frozen state, as well as during it, are not "normal"; they are designed to facilitate survival of freezing and are therefore specifically turned on at this time.

Regulatory Metabolic Functions before Freezing

On exposure to falling temperatures, cold-hardy insects do not activate the compensatory metabolic mechanisms often observed in ectotherms during cold acclimation. Rather, cold-hardy insects appear to use thermal change as a signal for triggering major alterations in metabolism to orchestrate measures of cryoprotection. Two of the most important cryoprotective measures activated at this time are the biosynthesis of nucleating agents and the biosynthesis of polyols. Unfortunately not much is known about the former, but the latter has been well described as a good example of how cold-hardy insects are able to take advantage of falling temperatures as regulatory signals for activating critical new metabolic functions. Thus in *Eurosta*, low temperature alters both the rates of, and the pathways used in, carbohydrate metabolism. Glycogen phosphorylase from cold-hardy insects is cold "activated" via an increase in phosphorylase *a* activity, a change leading to the degradation of glycogen and the accumulation of glycerol. In some species low temperature also appears to increase the percentage of participation of the pentose phosphate pathway, an adjustment that generates NADPH for polyol synthesis. Although cold acclimation has little effect on the general enzymatic makeup of the larvae, a few selected enzymes involved in polyol synthesis (phosphorylase, phosphofructokinase, and polyol and sorbitol dehydrogenases) show *increased* activities in *Eurosta* larvae acclimated to cold.

Although all the aforementioned processes are important, the main mechanism for regulating glycerol and sorbitol syntheses involves differential temperature effects on the kinetic and regulatory properties of glycolytic enzymes, particularly at the phosphofructokinase locus. Thus, as temperature decreases from 23° to 13°C, glycerol production in the larvae is facilitated because glycogenolysis is depressed less than are other processes. Whereas the flux through phosphofructokinase is maintained, carbon flow through pyruvate kinase appears to be inhibited (Figure 7.1). So, the net result is a diversion of triose phosphates away from terminal regions of the glycolytic pathways toward glycerol. At even lower temperatures, however, phosphofructokinase is itself inhibited. At 5°C the enzyme is only weakly activated by AMP, cellular concentrations of specific inhibitors (glycerol 3-P, sorbitol) are high, and the enzyme shows a high Q_{10} value. These effects result in reduced flux to the triose level, with sorbitol rather than glycerol synthesis being favored in consequence.

Why have a dual cryoprotective system? Why not rely only on glycerol or only on sorbitol? Some insights into this problem come from studies of *Eurosta* larvae. In the overwintering third instar, a dual cryoprotectant system leads to accumulations in the hemolymph of up to 650 and 170 m*M* of glycerol and sorbitol respectively. It turns out that the production of two polyols is important because each provides its own form of cryoprotection. Glycerol synthesis is anticipatory in the sense that it is stimulated by cooling in the range 25° to 5°C, well before first exposure to freezing conditions. The initial signal may be endogenous and released at the second to third instar transition in late summer. Once formed, glycerol provides a permanent level of cryoprotection over the winter, and rewarming does not stimulate glycerol degradation. Glycerol catabolism occurs only in early spring, when larvae enter a prepupal stage. Sorbitol production, in contrast, is not anticipatory but is directly stimulated by cold. Synthesis is activated by exposure to 5°C and continues down to temperatures as low as −8°C, the supercooling point. Sorbitol thus provides a

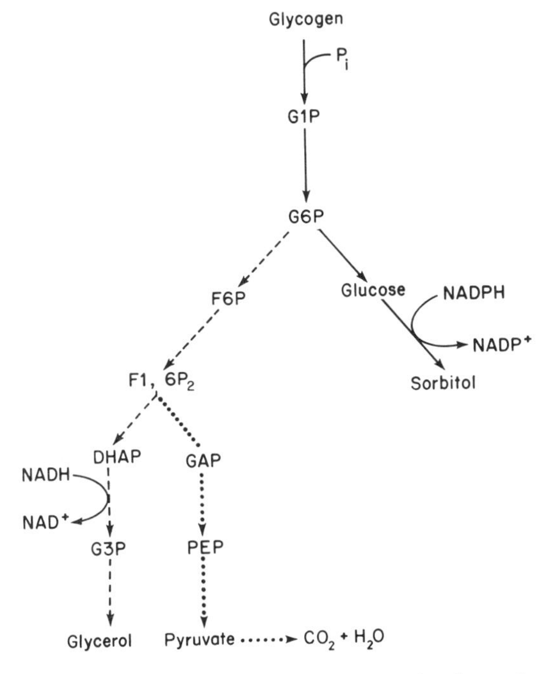

Figure 7.1 Metabolic pathways of glycerol and sorbitol synthesis in *Eurosta* larvae. Dashed and dotted lines show the pathways of carbohydrate degradation at cool temperatures (>5°C); solid lines show the route of carbohydrate degradation at temperatures below 5°C. Diversion of carbon flow into glycerol synthesis at higher temperatures probably results from inhibitory controls on glycolytic flux at the level of pyruvate kinase; at lower temperatures, inhibitory control at phosphofructokinase prevents further glycerol production and diverts carbon flow instead into sorbitol synthesis. After Storey (1983).

variable level of cryoprotection and is readily reconverted to glycogen in response to transient warming periods lasting more than two days.

Nucleating Agents

Whereas the formation of polyols requires regulatory modifications in a mainstream catabolic pathway (glycogenolysis), the formation of another class of compounds (nucleating agents) as part of a cryoprotective arsenal involves regulation of an anabolic process (protein biosynthesis). Although detailed studies of the biosynthesis of nucleating agents are thus far unavailable, these proteinaceous agents are known to be preferentially formed only seasonally (in preparation for overwintering). Their synthesis sites are not known, but it is known that they are polypeptides (>3,500 MW) that are transferred into the hemolymph prior to freezing. As in numerous other forms of insect diapause, freezing coincides with entry into dormancy, when DNA replication, cell division, RNA synthesis, and protein synthesis are gradually winding down. The polyols (glycerol and sorbitol) and nucleating agents formed during this phase of preparation for entrance into the frozen state, are in effect the last, and perhaps the most important, "end products" of metabolism formed before entrance into metabolic arrest. What are their functions?

Paradoxically perhaps, the role of nucleating agents is to minimize supercooling (or, in other words, to *raise* the temperature at which freezing of ECF occurs). The reason for this is clear if we recall that in the absence of nucleating agents, 0.9% NaCl will supercool to about −18°C. At that point, the spontaneous formation of ice can have disastrous effects on organisms. Rapid crystal growth punctures membranes; dehydration and osmotic imbalance of the intracellular space can occur and the potential is high for freezing to extend into the intracellular compartment. Under these conditions, preparatory metabolic processes are not possible. To avoid such problems cold-hardy insects induce a controlled synthesis of nucleating agents. Upon release into the hemolymph, these agents control formation of extracellular ice at subzero temperatures above which spontaneous nucleation would otherwise occur. All freeze-tolerant species rely upon this nucleation strategy and all show only moderate supercooling abilities, freezing occurring between −5° and −10°C. In summer, because of reduced amounts of nucleating agents in the hemolymph (or none at all), individuals of the same species supercool to much lower temperatures.

There is an interesting irony here: as part of the insect strategy for tolerating freezing, nucleating agents that in effect *facilitate* ice formation are produced. It is as if the insect is in a hurry to get it over with, since freezing may be inevitable in any event! The secret of this strategy, of course, is that nucleating agents in essence target extracellular water as the first (and it turns out, the only) freezing site allowed. As a result of nucleator action, not only is a controlled, slower freezing of the ECF compartment obtained at relatively higher subzero temperatures, but water is drawn out of cells to the limited extent required for an equilibrium to be reached, an equilibrium point at which the melting point of the concentrated intracellular fluid equals the actual temperature of the insect. In this condition, no supercooled compartment remains within the insect and the risk of injurious intracellular freezing is minimized. Even though the physicochemical mode of action of nucleating agents is not yet clear, our analysis adequately explains the reason why nucleating agents are formed by cold-hardy insects in the first place. A similar question must now be considered for polyols, the other main end products of pre-freezing metabolism.

Polyols and Intracellular Water

Reg Salt is to be given the credit for first working out, some three decades ago, how important glycerol is as a cryoprotectant in cold-hardy insects. Glycerol accumulates (sometimes up to 3 *M* levels) in the hemolymph of both freeze-sensitive and freeze-tolerant species. Alternative cryoprotectants are used by some species. Mannitol, sorbitol, erythritol, and threitol as well as several sugars (trehalose, glucose, fructose) can be formed and accumulated, either singly or in combination with glycerol or other cryoprotectants. Accumulations of proline or alanine or both (amino acids with possible cryoprotective functions) are also observed in overwintering insects.

Cryoprotectant polyols may have several functions, but in freeze-tolerant species the most important are (1) a colligative action in lowering the freezing point of intracellular water, (2) an effective increase in intracellular osmotic pressure to reduce cell dehydration during extracellular freezing, (3) protection and stabilization of protein, enzyme, and membrane structure and function, (4) modification of ice crystal growth in the extracellular environment, and (5) an increase in intracellular bound water content due to the hydrophilic nature of polyols. The last may yield the most reverberating effects (see later) and is particularly relevant to the theme of this book.

To put this problem into perspective we must reemphasize that when ECF freezes, the intracellular water:solute balance is perturbed. Ice forming in the ECF draws water out of cells and the concentrations of ions and other solutes inside must therefore increase. Dehydration and concentration of solutes past some critical level may disrupt metabolism and denature cell proteins and macromolecular complexes. To avoid or circumvent these potentially hazardous effects and further limit the chances of intracellular ice formation, cold-hardy insects apparently are able to adjust the nature of intracellular water. A similar process may indeed be utilized in anhydrobiosis, and the question how this can be achieved arises.

Most workers in this field now agree that water in cells can exist in more than one form. Three forms are widely accepted. (1) The water of *primary hydration* exists in intimate contact with the surface of solutes. This water is considered to be extremely firmly bound, because studies of hydration of proteins and nucleic acids show that it is not removable by desiccation and is not freezable (Kuntz and Kauzmann, 1974). (2) *Vicinal water,* or bound water, existing at greater (rarely specified) distances from solutes is ordered to a greater degree than solution water, because it is also influenced by its proximity to subcellular structures and solutes. Finally, (3) water in cells may also exist as *bulk water,* which presumably differs minimally, or not at all, from water in aqueous solutions (Beall, 1983; Storey, 1983). Currently available evidence indicates that during freezing in freeze-tolerant animals bulk water may have two fates: either it can move out of cells into growing ice crystals in the ECF or it can be "bound," in which event the fraction of intracellular bound water increases. Two important advantages accrue. First, not only does the ordered nature of bound water prevent it from freezing at the temperatures encountered, but the formation of bound water "shells" around subcellular macromolecules serves to protect them against denaturation, a function that polyols may take over in cases of greater desiccation (Crowe et al., 1983). Second, a large fractional increase in bound water may serve as a generalized "off" switch for much of cellular metabolism.

We can estimate the amount of bound water present in a system by measuring the rate of water loss during controlled drying of a sample over a desiccant. During drying of extracts of *Eurosta* larvae (in active stages or in frozen states), two phases of water loss are apparent (Figure 7.2). In an initial phase, water is lost at a high and constant rate—this is bulk water. During a second phase, water is lost at a constantly decreasing rate. Water lost at this slower rate is associated with subcellular macromolecules—the bound water. Such measurements applied to high-speed supernatant fractions from *Eurosta* larvae (Storey, 1983) indicate that low-temperature acclimation results in a significant increase in the bound water content. Bound water content increases by over 300%, from 0.19 g H_2O/g dry wt of total soluble components in larvae acclimated to 22°C to 0.63 g H_2O/g dry wt in larvae acclimated to −30°C, a value representing about one-third of the total water present (as bound water in the ICF plus ice in the ECF) in larvae at this temperature. The increase in bound water content of the soluble subcellular fraction is due to two factors, namely, to greater water binding by low-molecular-weight (dialyzable) and by high-molecular-weight (nondialyzable) fractions of the cell (Figure

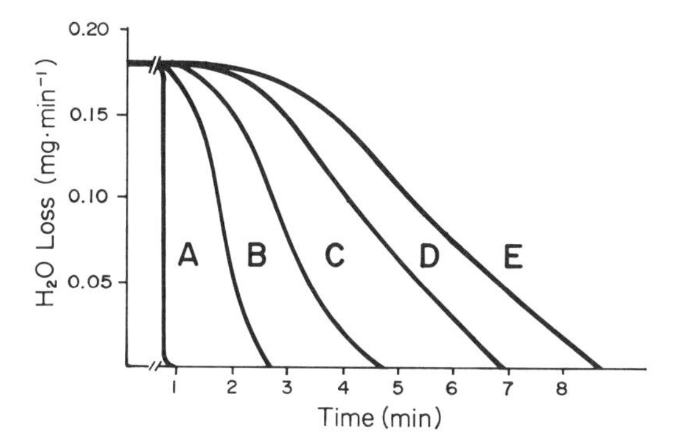

Figure 7.2 Measurement of bound water content. Samples are placed in a Knudsen cell inside the chamber of a Cahn electrobalance. After evacuation of the chamber to 90 torr, the rate of water loss from a sample into the surrounding desiccant is measured. After an initial phase at constant rate (representing the loss of bulk water), a second phase begins in which water is lost at a constantly decreasing rate. Bound water content is determined from the areas under the curves during this phase. The figure shows a schematic time-derivative plot of the rate of water loss from five samples of different bound water content: (A) homogenization buffer alone; (B) high-speed supernatant fraction from larvae acclimated to 22°C, dialyzed; (C) the same supernatant fraction, undialyzed; (D) supernatant fraction from larvae acclimated to −10°C; (E) supernatant fraction from larvae acclimated to −30°C. Modified from Storey (1983).

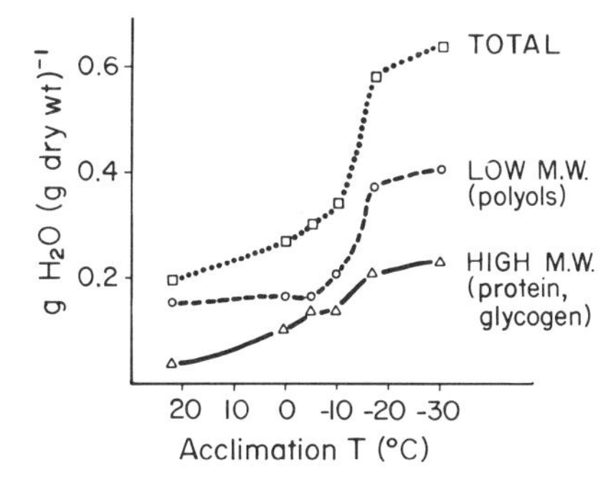

Figure 7.3 Effect of acclimation temperature on the amount of water bound by soluble subcellular components of *Eurosta* larvae. Total bound water is measured in undialyzed high-speed supernatant fractions; water bound by high-molecular-weight components is determined after dialysis of samples; water bound by low-molecular-weight components is determined by difference. Modified after Storey et al. (1981).

7.3). The increase in water bound by the low-molecular-weight fraction is mainly due to accumulations of high levels of glycerol and sorbitol by the larvae. An increased capacity for water binding by the high-molecular-weight fraction, on the other hand, probably results from physical alterations in soluble proteins (as the total amount of protein or glycoprotein in the larvae does not change with acclimation) or to increased availability of other high-molecular-weight components such as glycogen (Storey, 1983).

Effect of Bound Water on Metabolism in the Frozen State

Whatever the specific contributors to increased water binding in frozen insects may be, the effects on the metabolism seem fairly profound. The simplest mechanism influencing metabolism under these conditions may be the severe restriction of available bulk water. At overwintering temperatures, water in the frozen insects may exist in two principal forms: extracellular ice and intracellular bound water. This condition may be analogous to that in desiccated *Artemia* cysts, with the absence or presence of intracellular bulk water serving as a generalized "off" switch for cell metabolism (Chapter 9).

In addition, restricting bulk water availability in the cell may have more subtle effects on metabolic activity because of the effect of bound water on the accessibility of metabolites for enzyme reactions. By increasing the shell of bound water surrounding subcellular macromolecules (such as enzymes) at freezing temperatures, the organism has a way to localize intracellular metabolites into the bound water layer and thus to modify the ratio of free to bound metabolite pools in the cell. This process could critically influence the relative rates of different metabolic pathways simply by differential binding of metabolites.

Initial evidence along these lines is available from recent ^{31}P-nuclear magnetic resonance (NMR) spectroscopic monitoring of phosphorylated intermediates in larvae living either in frozen or in active states (Storey et al., 1985). These studies indicate that temperature change dramatically influences the ^{31}P-NMR spectra of the larvae as well as the apparent contents of phosphorylated compounds. As temperature falls, the relative intensities of peaks also fall, the decrease in the arginine phosphate and ATP signals being by far the most profound. This behavior is not observed in model solutions nor is it due to changes in the absolute concentrations of arginine phosphate and ATP in the larvae (as temperature change does not affect total arginine phosphate or ATP content measured enzymatically). The results appear to suggest therefore that decreasing temperature results in a progressive increase in the proportion of ATP and arginine phosphate molecules that are restricted in their rotational movement. Such a restriction may result from binding of these molecules to subcellular proteins, analogous to the "NMR-invisible" or undetectable ADP bound to actin (Meyer et al., 1982). The implication is that as the content of bound-versus-free water increases in cells as the temperature of the frozen organism falls, so also the ratio of bound to freely soluble high energy phosphates also seems to increase in the cold.

Just how general such shifts in the ratios of bound to free metabolites may be in the frozen state is not currently known. P_i translocations, for example, either do not occur at all or the shift is less dramatic. Nevertheless, it is clear that the large increase in the fraction of bound water with coincident decreases in bulk water may have a significant impact on metabolism in the frozen state. That is presumably why glycerol synthesis from glycogen (a pathway requiring ATP for the PFK reaction) ceases at the temperature at which free ATP signals are lost in NMR scans of frozen larvae—a temperature, incidentally, at which total extractable ATP pools are essentially normal! The implication therefore is that, unlike *Artemia* cysts and possibly other anhydrobiotic systems, where water loss during extreme desiccation serves at best as a kind of generalized "off" switch, cold-hardy insects seem to rely upon a subtle interplay between declining temperatures and increasing intracellular bound water to orchestrate a significant part of their metabolic responses for surviving freezing. It is as if they are exploiting the factors of freezing and using two of them (low temperature and bound water) as a kind of coarse control system, upon which more conventional regulatory mechanisms of catabolism and anabolism are superimposed. Thus it is not surprising that conventional allosteric regulation of metabolic enzymes (such as phosphorylase and phosphofructokinase) is retained under these conditions, and that the modulating role of H^+, so central in dormancy of *Artemia* cysts (Chapter 9), is used in frozen insects in a far more modest way.

Role of Intracellular H^+ in the Frozen State

In cells and tissues whose main intracellular H^+ buffers are imidazole based, it is now widely appreciated that pH_i in ectothermic animals may well vary with temperature with a slope of -0.016 to -0.018 pH units/°C. If this pattern prevails in any given system, it necessarily means that the fractional dissociation (α_{imid}) of intracellular imidazole groups must remain constant and this pattern, termed alphastat regulation, would hold for metabolites such as carnosine and anserine as well as for peptide-bound histidyls (Chapter 4).

In the case of frozen insect larvae, ^{31}P-NMR monitoring of ECF and ICF pH (by monitoring the highly H^+ sensitive chemical shift of inorganic phosphate in these two compartments) indicates that the ECF pH remains constant despite large changes in temperature. Intracellular pH, in contrast, increases with falling temperatures in the manner predicted by the alphastat hypothesis. For *Eurosta* larvae, the slope is −0.0185 pH units/°C over a much greater temperature range (+15° to −12°C) than is usually examined in other animals. As a function of temperature, then, change in pH_i in frozen insects is used as a means to maintain normal charge states (and therefore normal binding and catalytic functions) of metabolic enzymes and other proteins in widely varying thermal regimes. Alphastat regulation in frozen insects, as elsewhere, is a means of assuring normal catalytic and regulatory behavior, which otherwise could not be allowed due to thermal perturbation of imidazole dissociation states; and it thus represents a strategy for slowing down metabolism that is totally different from that used in anhydrobiotic systems (Chapter 9).

Summary

Freeze tolerance is a strategy utilized by polar- and temperate-zone terrestrial insects that are frequently exposed to temperatures below their freezing points. Surviving reversible freeze–thaw cycles is made possible by three biochemical adaptations. First, while remaining under allosteric regulation, cell metabolism is also differentially influenced by decreasing temperatures; in effect, cold is exploited to allow the preferential synthesis and accumulation of two cryoprotectants (glycerol and sorbitol) plus proteinaceous nucleating agents. Second, in prefreezing stages, nucleating agents are released into the hemolymph to raise the temperature at which freezing occurs and to facilitate slower, controlled freezing of extracellular water. An important fate of intracellular water is to move into extracellular ice, which creates the need for a third category of adaptation: protecting the intracellular metabolic machinery against freezing of intracellular water and against denaturation due to bulk water loss. By increasing the availability of polyols and of nondialyzable water-binding macromolecules, a greater fraction of intracellular water can exist as bound water. Because bound water is not subject to freezing at the temperatures encountered and because it protects macromolecules against denaturating, this adaptation simultaneously reduces the risk of freezing of intracellular water as well as denaturation of functionally important macromolecules. At temperatures in the range of −30°C, the metabolism of frozen insects is drastically depressed (to less than 1% of normothermic rates) and simplified, yet it remains under conventional allosteric regulating mechanisms. The requirement for both catalytic and regulatory functions, many of which depend upon constant fractional dissociation of imidazole groups at key binding sites, presumably explains why pH_i changes with temperature only modestly, enough only to maintain charge on imidazole groups thermally independent. This role of H^+ is a conventional one in ectothermic organisms and far less pivotal than is the H^+-mediated metabolic arrest in *Artemia* cysts.

Because of metabolic arrest, a molecule of substrate can sustain deeply frozen insects for over 100 times longer than normothermic ones. Biological time is in effect extended or slowed down by the same two or more orders of magnitude.

8

Frozen Frogs

The freezing tolerance of terrestrial insects living in temperate and polar regions is a capacity also found among other invertebrate classes, including other arthropods, helminths, and mollusks. Most readers, however, would not expect to find this capacity in any vertebrate class and may be surprised to learn that a few amphibian species are able to do just that, namely, to reversibly freeze (Schmid, 1982). In trying to understand how reversible freezing is accomplished and the selective pressures favoring its development, it is useful to recall that, in general, the distribution and activity of amphibians and reptiles are strongly influenced by temperature. In North America, for example, few reptiles and amphibians extend beyond the southern parts of Canada (Chapter 4). Among reptiles, the garter snake is actually the only species found across Canada at a distance as far north as James Bay, whereas in the Amphibia, a few anuran species (toads and frogs) range even further north; the range of *Rana sylvatica*, for example, extends beyond the Arctic circle.

Cold Adaptation Strategies

To overwinter successfully in such harsh environments, amphibians and reptiles could, in principle, harness the same cold adaptation strategies we have observed in insects: (1) avoidance (by migration or burrowing), (2) endurance (by supercooling), or (3) reversible freezing. Of these, most reptiles and amphibians choose the first alternative and avoid freezing by whatever mechanisms are available. Numerous turtles, newts, and frogs "hibernate" underwater (Chapter 4). The American toad burrows down a meter or more into the soil to avoid freezing temperatures. Salamanders usually hibernate using small mammal burrows, cavities around tree roots, or other natural crevices that reach below the frost line. Garter snakes assemble in large numbers in underground dens, where the temperatures are also well above freezing. In all these cases, avoidance is a necessary strategy of survival, as neither the second nor third of the preceding alternatives is possible.

The second mechanism, allowing for supercooling, is a common strategy used by insect species, but it is simply not accessible to vertebrates (Schmid, 1982; Storey, 1985). Measurements indicate supercooling points of only −2° to −3°C, not the −20° to −50°C supercooling points that characterize insects relying on this strategy. Why this should be so is not known.

The third choice, allowing for reversible freezing of extracellular water, is opted for by four species of anurans (Schmid, 1972; Storey and Storey, 1984): the spring peeper, the gray tree frog, the chorus frog, and the wood frog (*Rana sylvatica*). Although most of the biochemical information arises from studies of the wood frog, during overwintering all four species remain at the soil surface, well covered with detritus so that hibernaculum temperatures (and hence body temperatures) do not fall below −5° to −7°C. Because tolerance to freezing in these species is restricted to temperatures between −6° to −10°C, covering up with detritus is a required behavior for maintaining at least a modest margin of safety for the overwintering period. The margin between the temperature at which frogs freeze and the lethal temperature also is modest. Thus, extracellular freezing in *R. sylvatica* occurs between −2° and −3°C, within only 5°C or so of the lethal temperature. When exposed to freezing temperature, immature adults* rapidly supercool to −3°C; once initiated, freezing requires 2–3 hr or more for completion (Figure 8.1). Animals held at a constant −2°C also usually freeze, although some individuals may sustain a supercooled state for more than 24 hr. Frozen frogs have stiff limbs and opaque eyes and show no response to pinching. They do not breath, no heartbeat is observed, and no bleeding occurs when the aorta is severed. Large ice crystals occur in the abdominal cavity, beneath the skin, and interspersed within the leg muscles. Organs

*When collected in autumn, immature adults are animals that have metamorphosed from tadpoles during the previous summer. They have spent only a few months in adult form, weigh 1–2 g and are entering their first winter. Reproductive organs are not mature. Mature adults are those that have lived through one or more winter seasons. For *R. sylvatica*, size ranges from 5 to 8 g for males and 8 to 13 g for females.

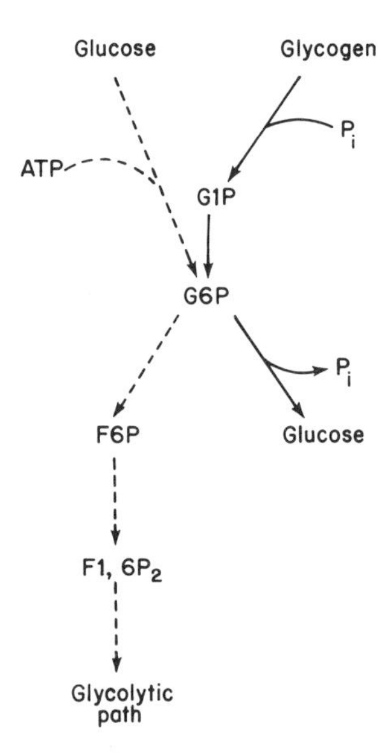

Figure 8.1 Pathway of glucose synthesis from glycogen in liver of *R. sylvatica*. Synthesis is controlled by activation of glycogen phosphorylase. Inhibitory control of hexokinase (HK) and phosphofructokinase (PFK) is needed to prevent the rephosphorylation of accumulated glucose. After Storey and Storey (1984).

such as the heart, liver, and stomach, however, even if surrounded by ice, do not actually freeze. About 35% of total body water is frozen in *Hyla versicolor* at −6°C, whereas estimates for *R. sylvatica* run as high as 48%. Such estimates seem consistent with the freezing of extracellular water only. Freezing of *H. versicolor* at −30°C results in an ice content of 58% and is lethal. Thus, estimates of ice content as well as visual observations of internal organs in frozen animals indicate that in frogs, as in other freeze-tolerant animals, only extracellular water freezes (Storey, 1985).

Regulatory Metabolic Functions Preparatory to Dormancy

Deposition of Fat and Glycogen Stores

Because many of the activities of temperate-zone amphibians are cued to seasons, it is not surprising that frogs show several fundamental metabolic adjustments in preparation for winter. For example, insulin turnover rates are adjusted so that plasma insulin levels reach their highest levels in the autumn; at the same time, steady-state glucagon levels decline to concentrations that are lower than at any other time of the year. As a result, conditions are suitable for the deposition of large reservoirs of glycogen and fat in the liver and fat body, respectively; the reserves will be required during the overwintering hypophagia (Storey, 1985; Farrar and Dupre, 1983).

Biosynthesis of Nucleating Agents

In insects about to enter an overwintering frozen state, preparatory metabolic functions include the biosynthesis, first, of nucleating agents and, second, of cryoprotective polyols. These products are needed because the initial freezing target in both groups of organisms must be the extracellular fraction of total body water. But extracellular freezing draws water out of cells, a process leading to cell shrinkage and to increased concentrations of metabolites and inorganic ions in the intracellular milieu. As a result of the increased concentrations of intracellular components, there is also an increased risk of denaturation of macromolecules and cellular organelles.

Unlike freeze-tolerant insects, frogs do not seem to rely on nucleating agents in blood and extracellular fluids. Several observations are consistent with this conclusion (Storey, 1985). In the first place, the supercooling capacity of whole animals is very limited. Second, apparently there are no differences in the supercooling capacities of *R. sylvatica* collected in autumn or spring. Third, if nucleating agents are used, the freezing point should be elevated; but, in contrast, freeze-intolerant frog species have only marginally lower supercooling points (on average 0.6°C lower) than do freeze-tolerant species. Indeed, this limited supercooling capacity of frogs may be the main reason why they do not need to rely on nucleating agents. This situation contrasts, of course, with that in insects, where freeze-intolerant species often have supercooling points of −20°C or lower. The summer forms of freeze-tolerant species also can supercool to almost this range, but during winter their supercooling point is raised into the −5° to −10°C range by the addition of nucleating agents to hemolymph. Why this difference should occur between freezing tolerance strategies in insects and in frogs is unknown, but it is quite clear that the freeze-tolerant amphibian is left with only the cryoprotective option.

Cryoprotectants

Although cold-hardy frogs differ from cold-hardy insects in not relying upon nucleating agents to raise the supercooling point, they are necessarily similar in relying upon intracellular low-molecular-weight cryoprotectants. Like many insects, *H. versicolor,* when exposed to freezing conditions, accumulates glycerol (0.3–0.4 *M*) in blood, tissues, and urine. This polyol plays its usual roles: increasing the fraction of bound versus free water and stabilizing macromolecules (proteins, in particular) and possibly membranes as well. Lower amounts of glycerol appear to be accumulated in immature adult stages. In both mature and immature forms, the pathway (glycogen→hexose phosphates→triose phosphates→glycerol) and its regulation are presumably the same as, or similar to, those found in insects. In immature adults of *H. versicolor,* however, the amounts of glycerol accumulated cannot fulfill the cryoprotective needs of freezing, so it is evident that other cryoprotectants must be made and accumulated. Interestingly, in immature *H. versicolor* glucose only makes up the cryoprotectant deficit, whereas in *R. sylvatica* glucose is the major cryoprotectant, with levels rising up to 0.55 *M* in the blood of freezing, exposed, adult females. Unlike insects, none of the freeze-tolerant anurans accumulate other cryoprotectant compounds such as sorbitol, fructose, or mannose; and even glycerol levels in *R. sylvatica* do not exceed about 1 μmol ml^{-1} (Storey and Storey, 1984, 1985).

Glucose in the Novel Role of Cryoprotectant

Cryoprotection is a novel function for a metabolite that we normally think of as being near the hub of intermediary metabolism. In other cold-hardy organisms, polyols are much more commonly observed cryoprotectants than are sugars, and there are some good reasons for this. In the first place, polyols are relatively inert chemically (in interactions with cellular molecules and structures) and biochemically (being formed in branch pathways off the mainstream of carbohydrate catabolism and not being readily catabolized by tissues). Glucose, in contrast, is a reducing sugar, and high concentrations raise the risk of nonspecific chemical interactions in the cell. It is also the preferred fuel for energy metabolism in many tissues, including brain, and plasma concentrations therefore are normally regulated by the actions of hormones such as insulin and glucagon. To use glucose as a cryoprotectant presumably requires major alterations in the metabolism of the frog, including the removal of pancreatic control over plasma glucose concentration and the removal of mechanisms preventing rapid oxidation of glucose or its recycling to glycogen. In view of the evident superiority of polyols over sugars as cryoprotectants, it is intriguing to consider the compromises being struck by cold-hardy frogs that rely upon glucose for this purpose. What advantages arising from glucose as cryoprotectant are being traded off at the price of minimal polyol contributions to freeze tolerance? The answer, it turns out, may well lie in the environmental signal cold-hardy anurans use to tell them when to activate cryoprotective measures.

We noted in Chapter 7 that regulatory metabolic functions in insects are anticipatory in the sense that in addition to endogenous hormonal clues the organism can take advantage of declining temperatures; as a result, prior to reaching the actual freezing point, cryoprotective measures are fully harnessed. This is not the situation in freeze-tolerant frogs. Frogs maintained at a constant 3°C, for example, regulate blood glucose levels at about 2.4 μmol ml^{-1}, as do high-termperature controls, even after three months of acclimation. The same result is obtained when frogs are held at 0°C; but plasma and tissue glucose levels are inevitably elevated after exposure to freezing temperatures. When the body temperatures of frogs are slowly lowered (by 0.5°C per day, for example) from 3° to -2.5°C individual frogs freeze at any temperature between about -1.5° and -2.5°C. In such cases, frozen frogs always are observed to have elevated glucose levels, whereas supercooled frogs, even if at the same temperature, always display normal (that is, low) plasma and tissue glucose concentrations. These observations indicate that cryoprotectant synthesis is signaled by the initiation of extracellular freezing per se rather than by subzero temperatures (Storey and Storey, 1984). What is needed, therefore, is a cryoprotectant that can be turned on and off at speeds faster than the rate of freezing. The narrow margin of error in anurans (compared with that in insects) provides the clue to the adaptive significance of using glucose, not polyols, as a cryoprotectant: as a mainstream metabolite, its production rates should readily outpace ECF freezing rates and so allow the organism to use ice formation per se as the signal for synthesis.

The expectation that freezing should lead to a relatively rapid accumulation of glucose is in fact observed in freeze-tolerant frogs. When maintained at -2.5°C, for example, blood glucose levels in-

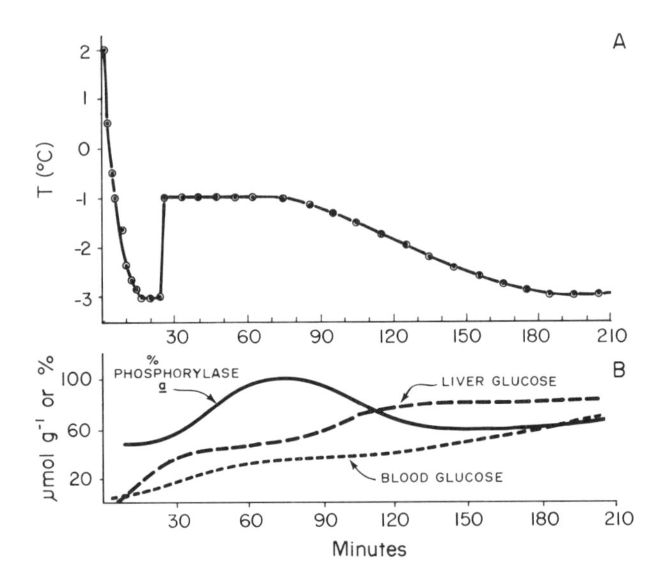

Figure 8.2 A. Typical pattern of freezing. The length of time during which individuals sustained the initial supercooled state at -3°C varied from 2 to 45 min. B. Composite freezing curves for *R. sylvatica*, correlated with concentration changes in blood and liver glucose and in liver phosphorylase *a* content over the course of freezing. Values for glucose are expressed in μmol ml^{-1} for blood and μmol g wet weight^{-1} for liver. Phosphorylase *a* activity in liver is shown as percentage of the total. Modified from Storey (1985).

crease to over 50 μmol ml^{-1} within the time required for fairly complete freezing (about 3 hr). When sensitive thermocouples are attached to the back of each experimental frog, it is possible to observe that during initial cooling to -3°C glucose levels in blood and liver remain unchanged (Figure 8.2). But, within 10 min of freezing initiation (indicated by the temperature jump to -1°C in Figure 8.2), glucose in blood rises to 16 μmol ml^{-1} and to about 40 μmol g^{-1} in liver. The rates of glucose formation in the liver represented by these data are in the range of 4 μmol g^{-1} min^{-1} at -1°C. Although it is not known what fraction of the ECF water freezes within this time period, the observed rate of glucose synthesis seems adequate for the cryoprotective needs. As freezing progresses under these conditions, glucose concentrations continue to rise toward an asymptote, presumably keeping cryoprotection closely tuned to the total amount of ECF water actually frozen. This pattern, interestingly enough, is represented repeatedly in all tissues and all fluid compartments thus far examined, a finding leading us to ask whether there is more than one site of glucose synthesis.

Liver, the Sole Tissue Site of Cryoprotectant Synthesis

The answer to the question of the tissue origins of glucose comes from a close analysis of measurements of glucose and glycogen in tissues of frogs exposed to freezing temperatures. As indicated in Table 8.1, glycogen levels in skeletal muscle, cardiac muscle, and kidney are low and do not change significantly when freezing is initiated, a situation that contrasts strikingly with that in the liver. As mentioned earlier, the liver in *R. sylvatica* stores huge quantities of glycogen, approaching 1 *M* levels in glucosyl units (Table 8.1), and this huge reservoir is apparently directly tapped for cryoprotectant synthesis. On exposure to freezing temperatures, liver glycogen is rapidly depleted to about one-eighth of starting amounts; in this way about 700 μmol glucose per gram of liver is made available for the rest of the animal's body. In a 10-g frog, this amount of glycogen-derived glucose could provide about 70 μmol g^{-1} throughout the body, assuming the liver to be 10% of the body weight. This whole-body value seems to be within the range of values actually observed.

Thermal Mediation of Liver Glycogen Mobilization

Glucose production can be relatively rapidly turned on as freezing begins because the pathway for its formation from glycogen is a sim-

Table 8.1 Concentrations of glucose and glycogen in tissues of control and low temperature-exposed *R. sylvatica*

	Glycogen[b]		Glucose[a]	
Sample	Control	Frozen	Control	Frozen
Liver	834	101	5.5	388
Leg muscle	37	38	1.3	27
Heart	48	44	2.1	198
Kidney	4	27	—	121
Blood	—	—	2.9	185

Source: Modified from Storey (1985).
a. Data are in μmol g wet weight^{-1} for tissues and μmol ml^{-1} for blood.
b. Glycogen is expressed in glucose units.

ple, three-enzyme sequence (Figure 8.1) that is low-temperature activated. Glycogen phosphorylase, the controlling enzyme, occurs in two forms, the active *a* form and the inactive *b* form. When frogs are exposed to freezing temperatures, the total phosphorylase activity in liver increases by over 500%, and the percentage of the enzyme in the active *a* form increases from 37% in control frogs to about 80% in freezing-exposed frogs. This result clearly indicates an activation of glycogen breakdown in the livers of frogs exposed to freezing temperatures. The correlation is very precise; Figure 8.2 shows that the percentage of phosphorylase in the *a* form increases rapidly with freezing and in parallel with the early rise in liver and blood glucose content. Muscle, however, shows only a small increase in phosphorylase activity (49%) with freezing exposure, and the percentage of phosphorylase *a* remains constant at about 35% in both control and freezing-exposed frogs. Thus, these data also suggest that the liver is the main site of cryoprotectant synthesis.

A comparable situation occurs in *H. versicolor*. Upon freezing exposure, the activity of total phosphorylase in the liver increases by about 200%, and the phosphorylase *a* content increases from about 40 to 60% (Storey and Storey, 1985).

Conversion of phosphorylase *b* to *a* is mediated by phosphorylation. Phosphorylase in the liver of mammals is typically under hormonal control: glucagon and adrenalin promote phosphorylation and enzyme activation through cAMP and/or Ca^{++} mediation, whereas insulin antagonizes phosphorylation. Some studies of cold-hardy insects also indicate that the enzyme may be directly activated by cold shock and that this may be a mechanism for phosphorylase activation during cryoprotectant synthesis.

Although these data on liver glycogen mobilization are satisfactory as far as they go, much remains to be done. There is still no information, for example, on how low temperature leads to rapid activation of phosphorylase. In some systems, low temperature may actually lead to phosphorylase inactivation; so this is by no means a trivial consideration. Additionally, the problem remains of how the normal, glucose-depressing action of insulin is circumvented to allow blood and tissue glucose levels to rise by as much as two orders of magnitude. However, the data are explicit in accounting for how the first step in glucose production from liver glycogen is turned on at the right time and at the right rate.

The next enzyme step in the pathway at which low-temperature adjustment seems to be required is the last in the pathway: that catalyzed by glucose-6-phosphatase (G6Pase). As in the case of glycogen phosphorylase in liver and in muscle, G6Pase activity expressed per gram wet weight of tissue appears to increase following exposure of cold-hardy frogs to freezing temperatures. In skeletal muscle of *R. sylvatica*, phosphorylase and G6Pase increase by about 50 and 70% respectively. It is important to emphasize that this is a differential, highly specific, low-temperature effect that is not caused by simple dehydration of tissue. This conclusion can be made with assurance, because fourteen other enzymes in muscle (functional in glycolysis, the Krebs cycle, the pentose cycle, gluconeogenesis, fatty acid oxidation, and amino acid metabolism) are known not to change in activity upon freezing of frogs. In liver, maximal activities of all enzymes that have been assayed increase upon freezing of frogs. However, the average increase for twelve of these enzymes (about 57%) is similar to an average increase in soluble protein content of about 55%; this finding indicates that freezing exposure causes dehydration of the liver by virtue of an outflow of water into ECF ice. In contrast to this dehydration effect, four enzymes in liver, all associated with glucose metabolism, show much larger percentage increases in maximal activities (Table 8.2). In addition to the 500% increase in total phosphorylase activity, G6Pase, glucose 6-phosphate dehydrogenase (G6PDH), and 6-phosphogluconate dehydrogenase (6PGDH) activities per gram

Table 8.2 Adjustments in enzyme activities during freezing exposure in tissues of immature adult *R. sylvatica*

Enzyme	Control[a]	Frozen[a]	Percentage increase
Liver			
Phosphorylase *a*	1.1	14.8	—
Total	3.0	18.6	520
Glucose-6-phosphatase	1.1	2.8	144
Glucose-6-P dehydrogenase	1.9	5.1	163
6-P-Gluconate dehydrogenase	1.4	3.4	141
Leg muscle			
Phosphorylase *a*	5.4	7.7	—
Total	15.0	22.3	49
Glucose-6-phosphatase	16.1	27.3	70

Source: Modified from Storey (1985).
a. Activities are in units per gram wet weight.

liver all increase by about 150%. In *H. versicolor,* there is an additional paradoxical 270% increase in hexokinase activity of the liver upon initiation of freezing.

Metabolic Control Problems in the Liver

From the available data, then, the metabolic situation in liver of cold-hardy frogs on initiation of freezing is summarized in Figure 8.1. To be able to generate a rapid, net production of glucose, the flow of carbon down the simple, linear path glycogen→G1P→G6P→glucose must not be significantly diverted toward the pentose phosphate cycle. How this is achieved is not known at this time. In fact, the low temperature-mediated increase in the catalytic activities per gram of tissue of G6PDH and 6PGDH is in itself perplexing; perhaps, as in cold-hardy insects, there is some need for reducing power in the form of NADPH during early phases of freezing, a requirement that would account for the preferential increase in capacity of this pathway. Nevertheless, it is clear that *most* glycogen carbon ends up as glucose, so in vivo the elevated catalytic capacities of these two dehydrogenases cannot be realized for long, if at all.

In all cold-hardy frogs, and particularly in the case of *H. versicolor,* an additional control of glucose metabolism must be imposed at the level of hexokinase. Again, paradoxically, the activity of this enzyme is actually elevated during freezing exposure. Along with the concurrent increase in G6Pase activity per gram of tissue, the stage is set for an energy-expensive cycling of carbon between G6P and glucose:

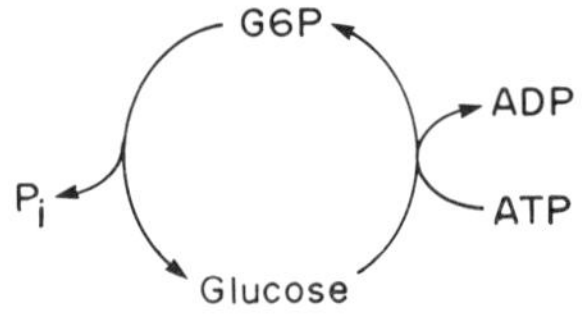

G6P

ADP

P_i

ATP

Glucose

Scheme 8.1

Although some cycling almost certainly must occur in this kind of metabolic situation, the available data clearly indicate that the net flux toward glucose is strongly favored in early phases of freezing. It is therefore implicit that hexokinase is under potent inhibitory control at this time. Glucose 6-phosphate at high concentrations, as may be expected during rapid glycogen mobilization, is well known as an inhibitor of liver hexokinases, and this may be the most likely candidate for this controlling function. Alternate mechanisms might involve adjustments in the availability of free ATP or in the bound:free ratio of hexokinase, because the latter is a classic amphibiquitous enzyme (presumably with two main intracellular locations: the mitochondria, bound to the external membrane, and the cytosol, where it may be involved in other binding interactions [see Chapter 2]). Finally, in this connection it should be emphasized that hexokinase inhibition is also requisite in other tissues in the body during exposure to freezing temperatures; otherwise, the glucose released at the liver would be further metabolized in these other tissues. The need for such a controlling function, however, becomes less severe once freezing is complete, because circulation at this time is fully blocked.

Profound Metabolic Arrest in Frozen Frogs

How much metabolism remains in frogs in the frozen state? The answer is, Not much. Frozen frogs do not breathe. If dissected while still frozen, in most individuals there is no heartbeat. A few frozen frogs sustain intermittent heartbeats, although hearts are pale in color and do nor appear to be blood-filled. Similarly the liver of frozen frogs is pale, but a large pool of blood, presumably frozen, is accumulated in the large vessels immediately above the heart. It is apparent, therefore, that during freezing oxygen and substrate delivery by the blood is fully blocked (Storey, 1985). The only other means of obtaining O_2 is via cutaneous diffusion. Thus a severely hypoxic and ischemic state prevails. Oxygen uptake is reduced to 1% or less of normal (K. B. Storey, personal communication), and the main energy metabolism remaining in the frozen state in both species so far studied in detail is the anaerobic fermentation of glycogen to lactate and the depletion of phosphagen and adenylate energy reserves. Over six days of exposure to freezing temperatures, lactate levels in tissues of *R. sylvatica* increase by 11.4, 12.8, 24.8, and 27.0 μmol g wet weight^{-1} in liver, leg muscle, heart, and kidney, respectively. Similar accumulations are found in *H. versicolor* and indicate rates of ATP turnover during this metabolically arrested state that are only 1/150 to 1/450 those of terrestrial animals at 15°C or 1/50 to 1/150 the rates at 0°C (Chapter 6). Nevertheless, fermentative energy production clearly continues; ATP production, however, does not quite keep pace with ATP consumption, so energy reserves necessarily drop somewhat with freezing exposure (Table 8.3). In *R. sylvatica* the creatine phosphate content of skeletal muscle drops by 50% during freezing but adenylate levels are maintained. Liver lacks a significant phosphagen pool, but freezing exposure results in a decrease in the total adenylate pool (50%) and a drop in energy charge from 0.79 to 0.51. *H. versicolor* shows similar patterns. Although it is not clear what sets the limits on freezing duration in cold-hardy frogs, the estimates of metabolic rates indicate that it certainly could not be fuel availability. A more likely possibility is that because of severe O_2 limitations the gradual accumulation of anaerobic end products to unacceptably high levels might place upper limits on the length of time these anurans can survive in the frozen state.

Summary

Of three major adaptation strategies commonly employed by cold-hardy organisms (nucleating agents to control extracellular ice formation, cryoprotectants for intracellular protection, and increased intracellular bound water content), freeze-tolerant frogs appear to utilize only the latter two. For reasons that may relate to using actual freezing of extracellular water as the signal for activating cryoprotective measures, frogs do not rely upon synthesis of nucleator substances with their subsequent release into the ECF. However, frogs share with invertebrates a reliance on low-molecular-weight carbohy-

Table 8.3 Levels of adenylates and phosphagen in tissues of control and frozen immature adult *R. sylvatica*

	Liver[a]		Leg muscle[a]	
	Control	Frozen	Control	Frozen
Creatine-P	0.31	0.07	13.1	6.65
Creatine	0.10	0.49	12.5	24.9
ATP	1.70	0.42	5.49	5.74
ADP	0.70	0.54	0.83	0.96
AMP	0.21	0.37	0.15	0.11
Total adenylates	2.58	1.32	6.46	6.86

Source: Modified from Storey (1985).
a. Concentations are in μmol g wet weight^{-1}.

drates as cryoprotectants and show a closely controlled set of molecular mechanisms for activating synthesis. Known enzyme adaptations for cryoprotection involve increased activities of a small number of key enzymes in glucose metabolism. Glycogen phosphorylase activity per gram liver and muscle is increased; in liver the fraction of the enzyme in the phosphorylated, or active *a* form, is increased. Along with relative increases in G6Pase, catalyzing the last step in the short pathway from glycogen to glucose, the net flux of carbon to glucose can keep pace with the rates of freezing of extracellular water, and thus fulfil most, if not all, of the organism's cryoprotective needs. Any remaining shortfall may arise from muscle glycogen, where the same pathway appears to be low-temperature facilitated. In addition to these specific adjustments allowing for controlled flux of glycogen→glucose, two additional enzyme loci must also be closely modulated during freezing. The first of these, G6PDH, competes for G6P with G6Pase at a branch point in the path from glycogen→glucose and must therefore be under inhibitory control to allow for large and relatively rapid net carbon fluxes to glucose. Similarly, hexokinase must be maintained under inhibitory control to prevent energetically wasteful cycling of glucose carbon back to G6P and to minimize the further metabolism of glucose. Finally, insulin regulation of blood glucose clearly is compromised during freezing, again by unknown mechanisms.

The metabolism remaining in frozen frogs cannot be based on blood-borne O_2 because the circulatory system freezes up along with the rest of the ECF. As far as is known, cutaneous respiration is the only source of O_2 for the low rates of oxidative metabolism still occurring, presumably at the periphery. Other energy needs are made up by modest rates of glycolysis estimated at 1% or less of normoxic ATP turnover rates. As in other metabolically arrested systems, then, frozen frogs do not make up energetic shortfalls arising from depressed oxidative metabolism by activating glycolysis. In effect, allowing metabolism to drift down as body temperatures drift down to below the supercooling point and by tolerating freezing, cold-hardy frogs extend biological time relative to clock time 100-fold or more.

9

Anhydrobiotes

Most cells and tissues are about 80% water, some a bit more, some a bit less; and this concentration of water (about 55 M) is usually maintained despite widely varying water concentrations in the outside medium. Terrestrial organisms and most higher marine organisms tend to lose water to the outside and must make up for the osmotic loss by increased intake, whereas in fresh water, aquatic organisms must resist the inward flow of water by appropriate organ-level adjustments. In many cases, then, and particularly in terrestrial environments, organisms must expend energy to prevent equilibrium from being attained between internal and external water. The reason for closely regulating intracellular water concentration, of course, is that it is the medium sustaining many, if not all, metabolically relevant reactions of living organisms.

There are several strategies available to organisms for "defending" the disequilibrium between water in inner and outer compartments. One possibility, usually the first line of defense in the face of water restriction, is to minimize water loss via usual channels. Desert mammals, for example, have developed kidneys with the ability to form highly concentrated urine; these may also lower metabolic rates (sometimes in estivation) to reduce respiratory water losses. Another strategy involves the development of water-impermeable coatings or barriers. Aboreal amphibians in dry, hot lands maintain on their skin a lipid-based coating designed as a barrier to outward water movement. Many desert arthropods also display special adaptations to minimize water loss across the cuticle as well as a means for absorbing water from the air. All such mechanisms have their own specific advantages, but they all share a common disadvantage—the high cost of maintaining water inside and outside compartments very far from equilibrium. Perhaps for that reason, some organisms when faced with severe water restriction problems simply give up the battle and lose intracellular water to the extreme limit of complete desiccation. This process necessarily requires that the organism enter into a dormant or encystment phase, because no normal life-supporting functions are possible if all intracellular water is lost. Nevertheless, the strategy is utilized (we are surprised with what high frequency), perhaps in part because it solves the problem of surviving the stress period and in part because it represents a large energy saving to the organism. In fact, the latter advantage may explain why some of the most capable anhydrobiotes (species capable of sustaining cycles of dehydration–rehydration) sometimes enter the requisite dormancy phase in response to secondary signals (and not necessarily simply in response to water lack). *Artemia salina* is one such species, and its cyst stage is also one of the best understood of dry biological systems. It thus represents an excellent place to begin our analysis of how cells can utilize the ultimate switch-off strategy and go ametabolic.

Signals Mediating Encystment in *Artemia*

Artemia is widely utilized in aquaculture efforts around the world, and because of this economic importance it is widely studied under laboratory conditions. Ecological studies, however, are not abundant, so the exact factors in the natural environment leading to encystment are not well known. Of possibilities such as salinity change, food lack, high temperature, or O_2 lack, it is now considered that O_2 limitation (or a generally more reducing condition in the external medium) supplies the critical environmental cue leading to encystment and dormancy (Persoone et al., 1980). We can well imagine why this should be so. Even if the adult *Artemia* is one of the best osmoregulators in the world, the process of maintaining appropriate disequilibrium states while living in a concentrated brine solution is evidently most energy expensive. With Na^+,K^+-ATPases and other ion pumps making a very large contribution to the organism's SMR, O_2 limitation would be expected to make the required rate of ATP generation unattainable by anaerobic metabolism. It stands to reason, therefore, that the organism activates an alternate strategy whenever O_2 supplies are threatened. The "encystment decision" appears to be made before O_2 actually runs out completely, because a distinct period of development (from fertilized egg to 4000-cell gastrula) is required for generating dormant cysts. Nevertheless, when external conditions are right, most (if not all) adult female *Artemia* must respond the same way: of two developmental paths available to them (one releasing swimming nauplii, the other releasing cysts), the latter is rather uniformly followed. That is why, periodically, whole salterns nearly fill up with huge quantities of miniscule brown particles (200–300 μm in diameter); these cysts are either observed floating at the water's surface or thrown by wind and waves onto the shore, where they may accumulate as a thick, meter-wide swath along the shoreline, often destined to fully dessicate in their wait for conditions favoring further development.

For *Artemia,* then, at least two classes of signals are needed to induce the anhydrobiotic state: the first is an exogenous category (necessary but not sufficient), acting upon adult females and setting the stage for secondary (or endogenous) signals, presumably derived from the female and mediating entrance of the developing embryo into an encysted or diapause state. It turns out that breaking the state of dormancy also requires exogenous and endogenous factors: a dehydration–rehydration cycle followed by a series of endogenous signals that mediate the return to, and completion of, development. Of the three processes (entrance into dormancy, maintenance of dormancy, and breaking of dormancy), entrance into dormancy is least understood. But because a great deal is known about the anhydrobiotic (cyst) stage and about breaking of diapause, a reasonable picture also emerges of the kinds of events that must underlie entrance into dormancy. These problems have been reviewed by Clegg and Crowe (1978) and by several authors in an excellent, three-volume series edited by Persoone et al. (1980).

Role of the Cyst Shell

One of the strategies available to organisms facing extreme environmental conditions (Chapter 1) involves "plugging the leaks" between the insides of cells or whole organisms and the outside medium. In its dormant cysts, *Artemia* has followed this strategy to nearly as extreme a degree as is possible. The process of building a protective barrier begins at the shell gland, which apparently sets up conditions permissive of development toward the cyst stage and which also begins to lay down a part of the tough coating or shell that the cyst requires. Of the two components to the shell, the outer one, or tertiary envelope, is synthesized and secreted by the shell gland, a process that begins soon after cleavage begins and is completed in about 36 hr. The embryo meanwhile lays down the inner layer (or embryonic cuticle), which includes the main permeability barrier, the so-called outer cuticular membrane. However, this layer is not formed until the shell gland has completed the outer tertiary envelope.

Once completed, the shell (which is fundamentally inert, containing no metabolically active components) is a powerful shield protecting the inner living compartment from the consequences of material exchange with the outside world. Thus, dormant cysts are impermeable to amino acids, sugars, nucleotides, glycerol, and a host of other organic compounds, and to essentially all ions that have been tested, including Na^+ and K^+. Interestingly, the cell membranes of dormant *Artemia* cysts retain no measurable Na^+,K^+-ATPase activity. The dormant cyst, in other words, has turned off one of the main drains on cellular energy metabolism and one of the main contributions to SMR in other organisms. About the only molecules that can permeate the dormant cyst are methanol, ethanol, O_2, CO_2, and H_2O. The first two of these are of no biological significance (and are probably chemical accidents arising from molecular similarities with H_2O). The ability to exchange H_2O, on the other hand, must be retained so that the cyst is able to fully rehydrate when appropriate ecological conditions are reestablished. Similarly, the permeability to O_2 and CO_2 must be retained to allow re-initiation of metabolism following rehydration. Aside from these few small molecules, the barrier to the outside world is remarkably complete.

The shell of the cyst, as part and parcel of a survival strategy that removes the intracellular milieu as much as possible from interaction and exchange with the external environment, renders the gastrula a "fortress *Artemia*": a living system so well protected that it is in effect isolated from most of the planet's chemistry.

Resistance to Extreme Physical Factors

A tough external coating may serve to protect sensitive intracellular components from damage due to permeant materials, but it cannot protect the internal milieu against physical environmental hazards. So it is all the more interesting that in addition to its protective barrier, the *Artemia* cyst seems to be well supplied with internal "defenses" against the outside world. For example, dried cysts appear to display no lower thermal tolerance limit; experimentally dormant cysts have survived prolonged exposures to −270.8°C. Although there is an upper limit (60°–70°C) to the temperatures that dormant dry cysts can tolerate essentially indefinitely, they tolerate temperatures up to 103°C for short (hour-long) periods. Similarly, dry cysts display no known lower pressure limit, about 10^{-8} mm Hg having been experimentally tested. (High pressure effects have not been studied to our knowledge; but judging from other properties of dried cysts, we would predict essential insensitivity to any values of absolute hydrostatic pressure ever to be found in the habitable portions of this globe.)

Many studies have emphasized the extreme radiation resistance of dried cysts, including resistance to X rays, γ rays, fast neutrons, proton beams, high-energy electrons, and ultraviolet radiation. Because of their hardiness, *Artemia* cysts have been used in space experiments. In these studies, cysts were housed on the outside of spacecraft and exposed directly to conditions of outer space. Little effect on viability was observed, except for cysts hit directly with cosmic heavy ions! "Fortress *Artemia*" is a concept, therefore, that extends beyond the cyst's outer covering. We are, in fact, dealing with an extremely resistant organism, whose cells are accurately described as among the most resistant of all animal cells to environmental hazards.

Water Substitutes and Desiccation

It is interesting that the outstanding resistance of *Artemia* cysts to various environmental factors depends upon one factor above all others—the ability to tolerate complete desiccation. Although not all biochemical and biophysical aspects of desiccation are fully understood, it is well known that to survive desiccation initial accumulations of polyols or polyhydroxy alcohols, such as glycerol and trehalose, are needed. Dormant *Artemia* cysts, for example, are about 4% glycerol by dry weight, and trehalose may constitute up to 14% of the organism's dry weight. There exists a strong correlation between the accumulation of these two polyols and survival of the desiccated state, and these compounds are formed during development of the dormant gastrula from endogenous glycogen reserves (Clegg and Crowe, 1978).

Polyols appear to have two functions in the desiccating organism. Their most basic role has been described as that of a "water substitute," for they may undergo hydrogen-bonded interactions with polar or charged groupings on cell organelles and macromolecules, thereby functionally "replacing" water in this particular role. In addition, polyols like glycerol may have the important effect of stabilizing protein structure at low water activities. Apparently this is achieved by the almost total exclusion of glycerol from the highly structured water surrounding proteins. Increasing the amount of glycerol in an aqueous solution of proteins thus leads to an increase in protein structural stability. The mechanism underlying this effect can be explained as follows: if a protein unfolds, a larger area of protein surface comes into contact with the solvent phase, a configuration leading to increased amounts of highly structured water. Because glycerol is excluded from this structured water, the addition of glycerol favors the compact, folded (native) structures of proteins. Thus, if the loss of cellular water during entry into the desiccated state leads to a destabilization of protein structure due, perhaps, to increased concentrations of low-molecular-weight species in the proteins' microenvironment, the addition of high concentrations of glycerol can be viewed as a strategy for keeping the proteins in native form during periods of desiccation.

Other Stabilization Mechanisms

The remarkable stability of intracellular macromolecules that under normal circumstances are easily inactivated has led to a search for stabilizing mechanisms in addition to those relying upon polyols. Two such additional mechanisms may operate in *Artemia* cysts: (1) packaging the potentially sensitive macromolecule into intracellular membrane structures, or (2) packaging them into organelles. The first mechanism is illustrated in the occurrence of a latent form of mRNA almost exclusively associated with intracellular membranes; this association is thought to serve both in stabilizing an inherently unstable macromolecule (mRNA) and in maintaining it in an inactive form (one not read, or indeed readable, by ribosomes). It is possible that protein components of these membranes are most important in the stabilizing function, although further work is necessary in this area to be certain. The second mechanism, which relies upon organelles for protecting potentially sensitive macromolecules against environmental factors, can be illustrated by yolk platelets, which appear to play a stabilizing function for associated enzymes. Glycogen granules may also serve as a medium onto which enzymes are plated out, thus contributing to their stabilization during desiccation and dormancy. Although the importance of these interactions is well appreciated, the mechanistic basis for stabilization is not well worked out; we mention them here mainly for the sake of completion.

Coordinated Transition to the Ametabolic State

In addition to the resilient properties of shell and contents, the functional properties of *Artemia* cysts have been carefully examined, and all the usual components of cell physiology and structure are known to be present (Persoone et al., 1980): the nucleus with a full complement of DNA, an active DNA polymerase, and low activities of DNAses; the key isozymes of RNA polymerases (for mRNA, tRNA, and rRNA synthesis); all the components of protein synthesis (mRNA, initiation, elongation, and termination factors, plus ribosomes); all the components of cellular energy metabolism (mitochondria, enzymes of oxidative metabolism, enzymes of anaerobic glycolysis, enzymes of lipid and protein catabolism, plus ample substrates for catabolism in the form of yolk platelets and glycogen, stored in granular form around the cytosol). Because *Artemia* cysts are able to reversibly dehydrate, the occurrence of what appears to be a normal intracellular organization is, of course, interesting, but by no means the most striking feature of dormant cysts; after all, cellular systems lacking these components would be eternally dormant. In contrast, what *is* of great biological significance is the orderly fashion in which all function has ground to a halt. The potential for DNA replication is present, yet DNA synthesis does not occur. The potential for cell division is present, but it too is not realized. DNA is not transcribed into ever-new classes of RNA. Although latent mRNA is present, it is not translated. Polyribosomes are not present, even though monoribosomes are abundant. Ribosomes (80 S) are devoid of most factors normally required for protein synthesis, although these factors are present in dormant cysts. Perhaps what is most surprising is that energy metabolism itself is fully halted or at least reduced to very low rates, despite the presence of a full enzyme complement and ample substrates. The cyst when fully desiccated is appropriately described as ametabolic, and one conceptual paradox reverberates through the entire literature: enzymes for all critical steps in cell function are present, as are their substrates; yet the reactions they are to catalyze do not proceed. Why not? Why are the enzymes maintained at all, for that matter; what possible role do they have in the ametabolic cyst? The answer to the latter question is that the enzymes and all key substrates must be maintained throughout the dormant period to assure later resumption of the living, metabolic state. It is, in essence, to be able to complete development that stable and potentially active enzymes must be retained right through the dormancy period. Although these are not needed in the ametabolic state, they are certainly critically important upon resumption of development.

What keeps the enzymes in a "turned-off" state, however, is another matter; and it now appears that two mechanisms are most widely utilized. One simply relies upon separating enzymes from their substrates; enzyme inactivity, in this case, can be viewed as a passive process (no substrate upon which to be catalytically active). A second mechanism, possibly more important and more generally utilized in depressing biochemical reaction rates in *Artemia* cysts, requires an active inhibitory regulation of key enzymes in cellular function. The magnitude and nature of the regulation problems involved may best be illustrated by an initial overview of events leading from normal metabolically active precursor cells to ametabolic, dormant ones.

Metabolism at Variable Water Content

Because in the natural state *Artemia* cysts typically undergo a desiccation–rehydration cycle before resuming normal development, the first possibility we must consider is that water lack per se is a sufficient (and necessary) signal for metabolic arrest. This question has been closely addressed by Clegg and his coworkers (Clegg, 1979, 1984), and the indications are that metabolic rates of *Artemia* cysts are remarkably independent of water over large changes in content. The actual effects on metabolism depend upon degree of desiccation. Between full hydration levels (1.4 g H_2O/g dry weight) and 0.8 g H_2O/g dry weight, metabolic rate is independent of cell water content, whereas between 0.65 and 0.8 g/g, metabolic rate varies directly with this parameter. As cells are desiccated below 0.65 g/g, the metabolism of macromolecules (proteins and nucleic acids, in particular) gradually becomes arrested; and by 0.3 g H_2O/g dry weight, only the metabolism of small molecules can be expressed (involving a few mainstream pathways such as glycolysis, amino transferases, and so forth). Cells cannot be dried below about 0.1 g/g dry weight, and, at this desiccation level, they are essentially fully arrested and ametabolic.

That metabolism stops when all water is lost is not too surprising. What is surprising is that cyst metabolism may be only mildly affected, even after loss of over 50% of the cellular water. Further comparative studies show, however, that this is not unique to *Artemia* cysts, but seems to be characteristic of animal cells in general. It is now known in cryobiology, for example, that animal cells may reversibly lose this amount of water without significant damage. Similarily, recent ultrastructural studies have shown that several mammalian cell lines in culture can be reversibly dehydrated with 2 *M* sorbitol in the medium, again with no serious ultrastructural damage. Studies of fibroblast-like L-cell cultures show that as much as 80% of total cell water can be removed (by incubation in sorbitol), a change amounting to a 65% decrease in cell volume with no loss of viability. This drastic change causes only a 10–15% decrease in glucose oxidation and even smaller perturbations of glucose carbon flow to other intermediates. At 65% of normal volume, shrunken L-cells contain only about 0.6 g H_2O/g dry weight, or a hydration level well within the range of many protein crystals. Yet their metabolism is almost the same as controls. Clegg (1984) takes these data to indicate extensive organization of enzymes concerned with metabolism in the so-called aqueous compartments of cells. He and others have persuasively argued that most (and possibly all) metabolic enzymes in effect are not "in solution," at least not in the "bulk water" or "soluble phase" of cells. In his view, strong selective pressures favor metabolic processes that have "escaped from water" or, more precisely, from simple, mass-action controlled, solution chemistry. Rather than a universal solvent for metabolic reactions, the aqueous phase per se is considered more a communication system allowing the rapid and efficient transfer of metabolites around the cell, while enzyme pathways per se are organized in cellular ultrastructure as semi-independent units, each with its own functional integrity. The implications of this hypothesis are vast and have been discussed elsewhere (Clegg 1984; Appendix B). For the moment it will suffice to emphasize that even though extreme dehydration undoubtly contributes to general metabolic arrest in *Artemia* cysts, in itself a lack of water seems to be too nonspecific a signal to control the orchestrated events occurring at this time (however, see Chapter 7, where this problem may be avoided). Thus we must look to more conventional regulatory processes.

Sites and Possible Regulatory Mechanisms during Entrance into Dormancy

At the outset, we must emphasize that although a lot is known about the controlled entrance into dormancy in *Artemia* cysts, knowledge of an unequivocal signal that kicks off the whole train of events is lacking. Nevertheless, a train of events is known to occur; it tends to feed back upon itself and to involve at least five levels of control during entrance into encysted dormancy. These are outlined separately below.

Control of DNA Replication and DNA Polymerase Activity

The most fundamental level at which cell activity is blocked in *precyst* stages is DNA replication and cell division. DNA polymerase is present in dormant cysts, but remains inactive. It is clearly in a controlled, inhibitory state, but the mechanisms maintaining that state are not known. DNase activity, important in replication, recombination, and DNA repair, is also maintained at low levels, again by unknown mechanisms. One model invokes control via acetylation of histones; histone acetyl transferase, a regulatory enzyme displaying sigmoidal saturation curves for its two substrates (acetyl CoA and histones) is a possible site of control, but how these properties are translated into, or contribute to, the block in DNA polymerase function is not known.

Transcriptional (DNA→RNA) Sites of Control

The next loci of control in the flow of information from DNA to proteins are to be found at the level of DNA transcription to RNA. The activities of three enzymes are of concern in this context: RNA polymerases I, II, and III. The transcription of DNA to ribosomal RNA is catalyzed by RNA polymerase I; during dormancy, this transcription occurs at about 1/10 the level found in "normal" cells later in development. RNA polymerase II catalyzes the transcription of specific DNA sequences to precursors of mRNA; during dormancy the activity of RNA polymerase II is of that typical of cells later in development. RNA polymerase III catalyzes the transcription of DNA to tRNA and 5 S RNA, and it too shows low activity during dormancy.

It is now known that the above controls are achieved in at least two ways. One kind of control simply leads to low enzyme production rates and gradually reduced concentrations of all three RNA poly-

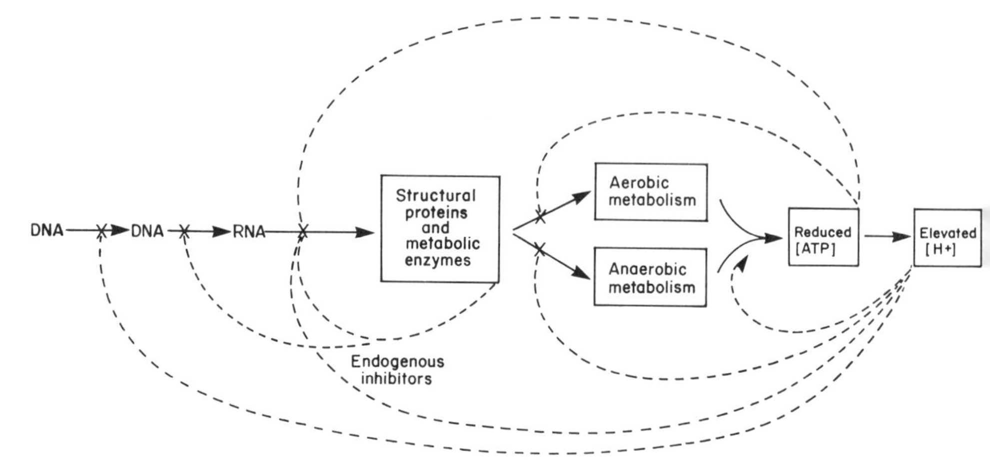

Figure 9.1 A diagrammatic summary of strategic self-reinforcing control circuits during entrance into dormancy in *Artemia* cysts. Details given in text.

merase isozymes. Because catalytic function is proportional to enzyme amount, this simple mechanism contributes to reduced rates of transcription (see Figure 9.1 and Persoone et al., 1980).

A second, finer level of control revolves around in vivo inhibitors that further reduce the catalytic activity of RNA polymerases. The mechanism of action is interesting. For RNA polymerases, a DNA template serves as a kind of substrate; and, in principle, an inhibitor could act either through effects on the DNA template or upon the RNA polymerase per se. Nature, it turns out, relies upon the former mechanism; thus, the more DNA, the less inhibition, with the percentage of inhibition being independent of RNA polymerase concentration. The effect of the in vivo inhibitor, therefore, is upon initiation of the RNA polymerase reaction; once the reaction is proceeding, the inhibitor is without effect and cannot influence the elongation of RNA chains. In cell economy terms, however, locating the inhibitor-sensitive site at the level of the DNA template is understandably efficient.

RNA→Protein Translational Control

A "fail-safe" strategy of regulated entry into dormancy is nowhere better illustrated than in the next steps in the control of enzyme content and composition. The situation here is extremely complex, but for convenience it can be categorized into five kinds of mechanisms.

Oligonucleotide control of translation. Probably the closest workers in this field have come to identifying the initial "on switch" that catalyzes the train of events leading to complete translational arrest is in the discovery of translation-blocking oligonucleotides. The inhibitor oligonucleotide is about 20 bases long, with a molecular weight of only about 6,000 (especially rich in uracil). Its formation from precursor RNA is catalyzed by a specific RNAase (T_1), and its action is highly specific in blocking chain initiation and peptide elongation. It is apparently nonspecific in the kinds of proteins whose synthesis is blocked, and its appearance during entry into dormancy would therefore be expected to lead to a gradual reduction in the levels of many enzymes (including RNA polymerases). Parenthetically, it may be noted that a reversal of this action may occur at the end of dormancy by the production of another oligonucleotide (formed from precursor RNA by another RNAase, the A enzyme) that can specifically bind the inhibitor and thus inactive it. Some such mechanism must exist to initiate reversal of dormancy and resumption of development when conditions are right.

In addition to the oligonucleotide modulator just described, a specific oligonucleotide inhibitor of poly(A)$^+$ RNA translation is known, although its long-range consequences are not yet clarified.

Finally, a lot of attention has been focused recently on the occurrence of extremely high concentrations of diguanosine tetraphosphate and triphosphate (G5′PPPP5′G and G5′PPP5′G). The interest derives from the well-known fact that close chemical analogues of these compounds (methylated at the N-7 position) are potent inhibitors of chain initiation; the methylated form of G5′PPP5′G is a particularly potent analogue of cap nucleotides and is therefore favored as a possible controlling element during entry into dormancy.

Translation-blocking proteins. Another category of control apparently operative during entry periods involves translation-blocking proteins. For example, protein chain initiation may be blocked by a protein kinase-catalyzed phosphorylation of eIF-2. The eIF-2 kinase could itself be under phosphorylation control by means of a cAMP-dependent protein kinase. Along these same lines, it is known that the cytoplasm and ribosomes of dormant cysts contain proteins that are inhibitory to chain elongation, and this too may contribute to a gradual slowing down of cellular activities. Artemin may be one such specific polypeptide associated with maintenance of dormancy.

Role of cytoplasmic membranes. A third category of control of translation depends upon specific interactions between informational macromolecules and cytoplasmic membranes. The best example of this in cysts is latent mRNA, much of which is associated with membranes. This latent mRNA is a product of nuclear genes transcribed during oogenesis or early in development. Its binding to membranes may serve two functions—stabilization and translational arrest—but how it is specifically mediated is not yet known.

Role of subunit–subunit complexing. Similar coupled functions (stabilization and functional inactivation) may be achieved through protein–protein complexing. For example, it is known that protein synthesis may be blocked at the enzymatic step responsible for binding aminoacyl-tRNAs to polysomes in the presence of GTP. This binding process is catalyzed by eucaryotic elongation factor T and is normally dependent upon at least three enzyme chains or subunits, all occurring in a freely dissociated form in solution. In the dormant cyst, on the other hand, at least two of these are tightly combined into a single complex that is functionally inactive. When protein synthesis resumes, the inactive complex can be dissociated again by GTP and aminoacyl tRNAs. It is commonly observed that liganded proteins are more stable to environmental extremes. Such improved structural stability in harsh environments may well be another advantage of protein–protein complexing during dormancy.

Role of ATP availability. Aside from effects arising from reduced rates of ATP cycling during dormancy, the simple parameter of reduced ATP concentration leads to rapid, if reversible, polyribosome disaggregation. As pointed out by Hultin and his coworkers (see Persoone et al., 1980), under conditions of low ATP concentrations, the net reaction

$$\text{polyribosomes} \rightleftharpoons \text{monoribosomes} + \text{mRNA}$$

is pushed far to the right. This may be an immediate cause of monoribosome accumulation in the dormant state and another important mechanism in blocking protein biosynthesis.

Control of Catabolic Processes

As shown above, information flow from DNA→RNA→protein is under inhibitory regulation at several levels in the overall process, an arrangement leading to reduced anabolic enzyme activities and reduced metabolic rates. Not surprisingly, the enzymes in most catabolic pathways thus far analyzed also are found to be present in low concentrations or to have low activities.

Mitochondrial oxidative metabolic pathways. The most graphic illustration of a repressed metabolic state comes from studies showing that mitochondria in dormant cysts are depleted in number and appear to be morphologically degenerate. Cristae are poorly formed and greatly reduced in abundance; and cristal surface area is drastically reduced. For this latter reason alone, we can assume that mitochondria are functionally inefficient. This impression is further supported by mitochondria that show a poorly staining matrix and express low activities of cytochrome oxidase and cytochromes *b* and *c*. Probably other components of respiration and phosphorylation also occur at low activities. Empirically, these effects are demonstrable with isolated mitochondria—as almost immeasurable respiration rates. In vivo these respiration rates may be even further reduced, because many (if not all) mitochondria are sequestered into yolk platelets, where they are presumably removed from sources of utilizable substrate (including oxygen?).

Anaerobic glycolysis. Whereas the shortfall in oxidative ATP generation would normally be made up by activation of anaerobic glycolysis (Pasteur effect), this does not occur in dormant cysts. Instead the paradox mentioned above is observed: flux is arrested despite the occurrence of requisite enzymes and substrates. Thus it is well established that both substrate supplies for glycolysis and enzyme titers are maintained throughout dormancy. Glycogen granules and abundant yolk platelets are evident in dormant cysts. At the same time, glycogen phosphorylase also is present, although it is sustained mainly in its low-activity, phosphorylase *b* form. Lactate dehydrogenase and presumably all the components of the glycolytic pathway are also present, again at reduced levels. Thus it is hard to avoid concluding that the enzymatic capacities for carbohydrate fermentation are well enough represented for some sustained function. Yet no lactate formation is observed even after five months of anoxia. In the extreme, then, there is literally no measurable glycolytic flux despite the maintained enzymatic pathway. To add to this paradox, intracellular pH appears to be drastically reduced (an observation to which we shall return below and which is usually associated with anaerobic glycolysis).

Metabolism of high-energy phosphate compounds. In the absence of oxidative and glycolytic sources of ATP, we might expect that dormant cells would turn to endogenous high-energy phosphate pools to sustain themselves. In most animal cells, this would be tantamount to relying upon internal phosphagens: in arthropods, the main phosphagen available is arginine phosphate. Interestingly enough, *Artemia* cysts do not contain any unusual supply of arginine phosphate, and it is not an ATP source during dormancy. G5′PPPP5′G, in addition to serving as a precursor for capping nucleotide analogues (which block polypeptide chain initiation), may serve as a source of GTP, and ultimately ATP, during dormancy. This process, augmented with endogenous ATP per se, appears to account for whatever low rates of metabolism are still required by anoxic dormant cysts.

The metabolism of other storage compounds. Finally, and largely for the sake of completing our thumbnail sketch of potential substrate sources for energy metabolism, it is important to add that, although the cell could in principle also harness storage proteins (vitellin, in particular) as a source of carbon for energy metabolism, this does not occur because of the induction of protease inhibitors. Similarly, it is observed that lipid mobilization is prevented during dormancy; but possible mechanisms here have not been examined.

Feedback and Feedforward Loops

What emerges from our brief analysis of a very exciting literature (see Persoone et al., 1980) is that *Artemia* embryo prepares for encystment by utilizing a "fail-safe" multistep strategy for gradually winding down both anabolic and catabolic machinery during desiccation. Although numerous feedback loops presumably must occur in the process, the most pivotal connections seem to be between protein biosynthesis and energy metabolism (Figure 9.1). As described schematically, the two processes (protein biosynthesis and ATP turnover) are in effect connected through mutually inhibitory control circuits: a reduced rate of protein biosynthesis leads to a reduced rate of energy metabolism and reduced ATP availability, which in turn leads to a reduced rate of protein biosynthesis, and so forth.

Although this two-way, mutually reinforcing, control circuit is easy to perceive as an elegeant way of slowing cyst anabolism and catabolism, the question of how such cycles of control get started in the first place remains unanswered. That is, we still need an exogenous signal to initiate the process. Let us be forthright in admitting at the outset that we do not know with certainty what that signal may be; in fact, we do not even know if there is only one signal. An intriguing possibility, however, is that the signal used is the smallest intermediate participating in metabolism—the H^+ ion.

Dormancy–Active Metabolism Transitions

To properly appreciate H^+ influences on these control systems, it is useful to recall that the *Artemia* embryo is released from its maternal ovisac in a profoundly depressed metabolic state (termed *aerobic* dormancy) that may last for years, if necessary. When cysts are desiccated and then rehydrated, development is reinitiated; but when they are rehydrated under *anoxic* conditions, dormancy can continue for at least five months (and probably indefinitely) despite rehydration. It has been possible to take advantage of this reversibility of the dormancy–active metabolism transition in *Artemia* cysts to show that the transition from aerobic development *into* anaerobic dormancy involves an acidification of over 1 pH unit, from about pH 7.9 to 6.3. Intracellular pH (pH_i) is effectively independent of external pH (pH_o), presumably as a result of the well-established impermeability of the cyst shell to counterions. In contrast, when metabolism and development of anaerobic dormant cysts are reinitiated by oxygenation, pH_i returns to about 7.9, a change representing the largest pH_i shift known to occur in any cell system under quasi-physiological conditions (Busa, 1985).

Although direct measurements of pH_i in *aerobic* dormant cysts are only now being attempted, preliminary studies provide circumstantial evidence that in these, as in *anaerobic* dormant cysts, pH_i is greatly depressed—and further indicate that low pH_i is a fundamental regulator of aerobic dormancy. Aerobic dormant cysts incubated in 40 m*M* NH_4Cl (pH_o 8.5) hatch as well as those conventionally activated (by desiccation–rehydration cycles) and better than untreated aerobic dormant controls. ^{31}P-NMR studies of anaerobic dormant embryos demonstrate that this level of ammonia increases pH_i by about 0.9 pH units. Conversely, when pH_i in activated cysts is depressed from the normal value of $\geq$ 7.9 to 6.75 (by incubation in 60% CO_2/40% O_2), respiration and development are reversibly arrested. Thus, treatment with CO_2 depresses pH_i and inhibits metabolism and development, imposing on these embryos a state closely resembling aerobic dormancy. Weak base treatment, on the other hand, raises pH_i and terminates true aerobic dormancy.

From parallel studies with other preparations (Busa and Nuccitelli;

1984), it is now evident that inhibitory H^+ effects could cut in at various levels in the complex process of dormancy (Figure 9.1). The most fundamental level of H^+ inhibition may well be at DNA copying and cell division, because both DNA polymerases and cell division are known to be reduced at low pH_i values. Proton inhibition of protein synthesis may represent the next general level of arrest; and finally, numerous steps in energy metabolism are known to be inhibited by low pH. In carbohydrate catabolism, phosphofructokinase (PFK) is usually considered to be most sensitive to H^+. Thus, although identification of specific sites of H^+ action are still under investigation (Busa, 1985), it is more and more widely appreciated that change in $[H^+]$ may be the most suitable general signal kicking off the process

$$\text{metabolically activated state} \rightleftharpoons \text{dormant state}$$

in either a forward or a backward direction, depending on the direction and magnitude of change in $[H^+]$. Because the system represented by the *Artemia* cyst is so closed, these pH_i transitions must be largely determined by internal metabolic events rather than by ion exchanges or transport at the cyst–environment interface. For example, from known changes in concentrations of G5′PPPP5′G and ATP, it is evident that significant acidification could be readily achieved by net ATP hydrolysis (which at around pH 8 yields one H^+/ATP hydrolyzed), and a similar degree of alkalinization could be achieved by net replenishment of the nucleotide pools (Busa and Crowe, 1983).

The final question is why H^+ was chosen (if it was) as a regulatory agent in the first place. The answer appears to center on an utterly unique feature suiting it for this role: between 10^{-6} and 10^{-8} *M*, H^+ is in perpetual equilibrium with H_2O, the solvent of life. Because water spontaneously ionizes, H^+ cannot be excluded from the intracellular milieu, even at the low water contents typical of *Artemia* cysts. Probably from the very beginnings of life, cell biochemistry has involved weakly ionized compounds and has relied upon acid–base catalytic mechanisms. Many protein functions have direct or indirect dependence upon imidazole groups, often located within binding (catalytic or regulatory) sites and expressing pK_a values close to pH_i values. Because the ionization states of such groups are exquisitely sensitive to modest pH_i changes within the physiological range, it is readily appreciated that numerous protein functions are similarily sensitive to pH_i changes within the same range. Intracellular H^+ concentration thus provides a powerful general signal for controlling metabolism without requiring the evolution of special receptor molecules. That may in itself supply adequate explanation for why pH_i seems to be a regulatory agent of choice for transitions between active and dormant stages in *Artemia* cysts. Choosing H^+ for this role, however, has one additional advantage: its obvious potential to communicate information regarding cell energy balance to enzymes and intracellular organelles that share no other common effector. This is an integrative regulatory property of pH_i that is clearly advantageous for this kind of generalized effector of metabolism.

Anhydrobiosis in Other Taxa

Extreme though its capacities are, the *Artemia* cyst is not alone in this world, for the ability to survive loss of all cellular water (save perhaps that bound very tightly to macromolecules) without irreversible damage is expressed in nearly all the major taxa (Crowe, 1971; Hinton, 1968). Keilin (1959) coined the term *cryptobiosis* ("hidden life"), but later workers prefer the term *anhydrobiosis* ("life without water") to describe this phenomenon. Two readily distinguishable categories of anhydrobiosis are recognized:

1. Organisms capable of anhydrobiosis only in their early developmental stages; for example, the seeds of plants, bacterial and fungal spores, the eggs and early embryos of certain insects. The *Artemia* cyst is the best-studied representative of this category.
2. Organisms capable of anhydrobiosis during any stage of their life histories; the best-known examples include protozoans, rotifers, nematodes, and tardigrades. Nematodes are the best understood in this category.

Soil nematodes when active are roundworms, 0.5–3 mm in length, and lacking circulatory and respiratory systems; water and gas movements are by direct diffusion through the cuticle. Although they live in the ground, soil nematodes are aquatic animals and require a water film around soil particles for normal activities. As quantitative techniques for extraction of dried nematodes from dried desert soils have been developed, it has become apparent that anhydrobiosis among nematodes is much more widespread than previously thought. Furthermore, anhydrobiosis is not confined to any particular life stage or trophic group of nematodes, adults and larvae of many genera being found when dried soil is rehydrated (Clegg and Crowe, 1978).

The stimulus for entrance into an anhydrobiotic state in nematodes, as in other invertebrates, seems to be water availability. When the animals are removed from water, a series of ordered morphological changes are initiated within the first 24 hours and are virtually completed within 72 hours. The whole animal undergoes a longitudinal contraction and coiling; intracellular organelles such as muscle filaments and membrane systems undergo ordered packing and change. Although a reduced water availability is a necessary condition for initiating these changes, it is not a sufficient one, because, on balance, Crowe and his colleagues conclude, all the above changes are regulated by the animal itself.

Although data on metabolic events at different states of hydration, comparable with those for *Artemia* cysts, are not available for nematodes, some studies report that desiccated nematodes do not consume O_2. Moreover, three decades ago Becquerel was able to revive nematodes, rotifers, and tardigrades after they had been exposed to temperatures as low as 0.05°K. He calculated that metabolic processes, if they occur at all, must proceed at only about 10^{-7} the rate of normally metabolizing specimens (Clegg and Crowe, 1978). Hinton (1968) and Clegg (1979) argue that this is not metabolism in the normal sense in which that term is used. Thus we assume that in nematodes as in *Artemia* cysts metabolism is at a standstill during anhydrobiosis. Many of the biochemical adaptations allowing reversible entry into this state in nematodes indeed seem to be similar to those described in *Artemia* cysts and will not be reviewed in detail.

Are Desiccated Anhydrobiotes Alive?

Are organisms such as dry *Artemia* cysts, dry nematodes, or dry rotifers alive in any meaningful sense? Keilin defined this condition as "the state of an organism when it shows no visible signs of life and when its metabolism is hardly measurable, or comes reversibly to a standstill." The key issue of this or any definition of anhydrobiosis revolves around metabolic activity, but in Keilin's definition it is evident that there is a qualitative difference between an organism in which metabolism is "hardly measurable" and one in which metabolism "comes reversibly to a standstill." In the former instance the organism is alive but very dormant, with the difference between dormancy and more active stages of such an organism being largely quantitative. In the latter ametabolic case, however, the organism cannot be considered "alive" by direct inspection, because in the absence of metabolism there is simply no way to know, at least by usual functional criteria. That is why for some workers in this field, Crowe for example, endogenous control over morphology is viewed as a key to this problem: so long as the structural integrity of the organism is preserved and remains intact, the organism is alive. It is dead when and if that integrity is destroyed. Although this principle (which in implicit form can be traced back at least to Hickernell in 1917) as stated applies primarily to higher levels of organization, it is equally applicable to the biochemical level. At this level of organization, structural integrity must be maintained by assuring an appropriate environment to each subcellular compartment, each cell, and each tissue; indeed, the adaptation of metabolic machinery allowing the establishment of a proper intracellular milieu for entrance into the ametabolic state may be viewed as the hallmark of this elegant strategy for surviving the total absence of water. Interestingly, only one other condition—adaptation for freezing tolerance—comes even close to this degree of metabolic arrest, with values for time extension approaching infinity.

10

Perspectives

Time. We and all other organisms live in it, yet surprisingly few biologists think about time in the context of control. When most biologists talk of temporal organization of biological systems, they usually focus on how biological activities can be coordinated with astronomical or clock time; that is, how circumannual, seasonal, lunar, or diurnal cycles and rhythms work, and how development sequences are controlled. Other scientists, including us, look at time and biology in a different way. Our view emphasizes that biological time and astronomical time need not always march in harmony, for the latter is unchanging while the former varies with rates of metabolic reactions. This relationship is most quantitatively expressed in the scaling of metabolism to body mass in animals. As elegantly explained by others (Chapter 1), biological or metabolic time varies inversely with metabolic rate and, since mass-specific $\dot{V}_{O_2}$ declines systematically with body mass for any group of similar organisms, biological time is systematically extended or slowed down by the same degree. That is why any given process in an animal the size of a mouse proceeds about 20 times faster than does the same process in an animal as large as an elephant. A second to a mouse means one-third of a minute to an elephant; a minute to a mouse is equivalent to about an hour for a large whale (see Table 1.1). Being big has its problems, and one universal component used in resolving them involves slowing down metabolism or extending biological time. In the previous chapters we argue that the same strategy is used to help circumvent problems arising from limiting availabilities of O_2, of heat, and of water. In all three cases, time is slowed down by activation of various metabolic arresting mechanisms and, in the extreme (in the case of anhydrobiosis), biological time appears to be fully halted; the organism becomes ametabolic and for practical purposes successfully escapes from time altogether.

Not surprisingly, the mechanisms used to achieve metabolic arrest depend upon the degree of time extension required (Table 10.1). However, in all cases, switching down means harnessing defense mechanisms for the controlled suppression of cell functions coupled with the controlled stabilization of cell structures (enzyme complexes, membranes, mRNA, organelles, and so on) both at cell–environment boundaries and within cells themselves. In perhaps one of the most important examples of metabolic arrest as a defense strategy—that of facultative anaerobes exposed to O_2 lack—hypoxia tolerance depends upon channel blockade in plasma, mitochondrial, and SR membranes, to avoid decoupling of metabolic and membrane functions as metabolism is suppressed. With stabilized membrane functions assured, metabolic arrest can proceed in an orderly fashion. Its early phases require the slowing down of oxidative metabolism (potentially achievable by O_2 conformity and by the regulated translocation of mitochondrial adenylates into the cytosol); in later phases, the slowing down of anaerobic pathways is achieved by a reversed Pasteur effect. In this hypoxia-adaptation case, tolerance to the environmen-

Table 10.1 Strategic steps in metabolic suppression processes

Step	Illustrative examples
1. Balancing membrane channel and membrane pump functions; membrane-boundary adjustments in order to minimize the cost of ion-specific pump fluxes	All cases of metabolic arrest
2. Suppression of oxidative metabolism (by O_2 conformity and mitochondrial adenylate translocation)	Hypoxia tolerance, torpor, estivation, freezing tolerance, anhydrobiosis
3. Suppression of glycolytic activation (by reversed Pasteur effect)	Hypoxia tolerance, anhydrobiosis
4. Arrest of protein synthesis	Freezing tolerance, torpor, anhydrobiosis
5. Arrest of RNA synthesis	Torpor, freezing tolerance, anhydrobiosis
6. Developmental arrest; arrest of DNA replication, cell division, and differentiation processes	Torpor, estivation, anhydrobiosis

tal insult seems to depend upon a simple two-pronged mechanism (that is, hypoxia tolerance = metabolic arrest + channel arrest), with seemingly little fundamental adjustment required elsewhere.

In contrast, as the degree of time extension required to survive the stress rises, the mechanisms utilized are understandably more complex, gradually extending from the cell boundary, through cell metabolism, all the way to the genetic level (Table 10.1). In the extreme case of anhydrobiotic organisms, for example, switching off requires extensive defense measures at the organism–environment boundary as well as at numerous intracellular levels of organization. Not only must aerobic and anaerobic metabolism be brought to a gradual and complete halt, so must all the machinery of translation and transcription, thereby leaving the organism morphologically "alive" but ametabolic. Even in this extreme case, the principle of arrested functions being coupled with stabilized (protected) structures is not violated; indeed, it may be most clearly expressed.

Interplay between Clock Time and Adaptational Strategies

Generally, there is a close interplay between clock time and biochemical adaptation strategies "chosen" by organisms in specific environ-

Table 10.2 Interplay between clock time and adjustment strategies

Clock or astronomical time available for adaptation	Adjustment strategy	Adjustment mechanisms
Seconds to minutes (infinitely small fraction of life cycle)	Acute	Built-in "machinery" adjustments; structure–function adjustments in enzymes and transport proteins; metabolite flux adjustments
Hours to days (fraction of life cycle)	Acclimatory	Biosynthetic adjustments; alterations in storage substrates, in enzyme concentration, in enzyme kind, in membrane and organelle structure
Phylogenetic (many life cycles)	Adaptational	Genetically fixed, profound structural and functional adaptations; adjustments in fermentation pathways; channel density and other membrane adjustments; specialized control circuitry

mental conditions. These can be termed acute, acclimatory, or adaptational responses, respectively, with boundaries defined in terms of the generation time of each organism (Table 10.2). By using animal extremists in this book to illustrate many of our arguments, we have intentionally emphasized the third category above, or responses that clearly have developed over phylogenetic history and that are genetically based. It is important to remember, however, that arrest-type defense mechanisms must be harnessable on a moment-to-moment basis as well. This may be well illustrated by short-term responses to hypoxia. Recent studies by Andersson and Jones (1985), for example, indicate a rapid channel blockade in mitochondrial membranes as a standard defense against hypoxia, a strategy resulting in the maintenance of K^+ and H^+ concentration gradients across the mitochondrial inner membrane at least for short time periods in the absence of oxidative metabolism. It is as if ion-specific transmembrane proteins in mitochondria are lockable in "off" conformations as an instantaneous defense against O_2 lack (see also Jones et al., 1985). Such structure–function adjustments may be the most rapid of the acute responses available to cells, because the environmental stress is also the signal for adaptive adjustments.

Endogenous Arresting Mechanisms

More complicated acute responses to environmental stresses involve more complex signal transduction mechanisms. For example, because of the close correlation between metabolic arrest capacities and hypoxia tolerance, Epstein et al. (1985) suggested that *endogenous* arresting agents may play an important role in modulating the susceptibility of tissues to anoxic injury on a short-term basis. Possible candidates for such roles are adenosine (perhaps along with other purine metabolites) and derivatives of arachidonate. Although the modes of action of these compounds are not understood in detail, adenosine is known to be derived from intracellular adenylates under conditions in which ATP hydrolysis rates exceed synthesis rates. AMP concentrations increase as a result, and some AMP is thought to cross the plasma membrane, where it is hydrolyzed by the ectoenzyme, 5′-nucleotidase. At low concentrations adenosine is a potent vasodilator and also activates inhibitory receptors that retard adenylate cyclase activity and thus suppress metabolism in excitable and secretory tissues. Presumably via such mechanisms, at concentrations of 10^{-8} to 10^{-5} *M*, adenosine substantially reduces ion transport-linked O_2 uptake of the thick ascending limb of the mammalian kidney (Epstein et al., 1985) and thus may constitute a metabolic arresting mechanism for the endogenous protection of the organ against injury during times of inadequate O_2 availability.

The mode of action of arachidonate derivatives, such as prostaglandin E_2 (PGE_2), is even less well analyzed. However, these compounds are known to be produced by cells such as those of the medullary thick ascending limb (mTAL) cells in perfused kidneys; PGE_2 in particular is known to profoundly decrease transport-associated O_2 consumption. Epstein et al. (1985) argue that these and a variety of other mechanisms (some of which are yet to be discovered) serve as short-term mechanisms permitting hypoxia-sensitive cells to function safely in microenvironments with unreliable O_2 supplies. Presumably, the reason such mechanisms had to be developed in the first place is because periodic O_2 limitation is an ever-present hazard to *all* aerobic organisms; therefore, rapidly activable defense mechanisms must be available to even the most O_2-dependent living systems.

In principle, exactly the same considerations should hold for environmental parameters other than O_2 availability, although these have not been explored as extensively. Margules (1979), for example, proposes that a regulated adjustment in the balance between β-endorphin and an endoloxone-type substance (the former favoring substrate conservation and metabolic suppression, the latter reversing these effects) may serve as an endogenous mechanism for switching down metabolism during mammalian hibernation. Something along these lines—for example, a genetically fixed and modified balance between two such antagonistic systems—may well explain why some species are permanently hypometabolic (for example, the slow loris among mammals [Whittow et al., 1977] and deep-living pelagic teleosts among fishes [Childress and Somero, 1979]).

Advantages of Metabolic Arrest

The ultimate benefit of metabolic arrest as a response to O_2, heat, or water deficiencies, of course, is survival of the organism despite the stress. Perhaps that goes without saying. Yet it is instructive to reemphasize (Table 10.3) that enhanced survival is a direct outcome of only two categories of advantage. Probably the main advantage of the arrested state is that it reduces the effective length of the stress period in proportion to the degree of metabolic arrest. Because freshwater turtles in northern ponds activate powerful metabolic arrest mechanisms during underwater hibernation, an overwintering (180-day) "dive" is metabolically equivalent to a 1-day episode at normoxic ATP turnover rates at 20°C. For a diving mammal operating at a constant and high temperature, a 60-min period of hypoxia sustained by a hypoperfused kidney may, because of metabolic arrest mechanisms, be equivalent to a 1-min hypoxic stress for the undefended organ at normoxic ATP turnover rates. In both kinds of systems, the metabolic problems of O_2 lack (depletion of fermentable substrate and self-pollution by anaerobic end products) are proportionately minimized. Exactly the same principles apply to metabolic arrest as a mechanism for solving problems of low temperature and of low water availability. *Eurosta* larvae, for example, may remain frozen for perhaps 120 days during harsh winters. But for them this time is equivalent to about 1 day of hypophogia at normothermic ATP turnover. The size of metabolic problems arising and the ease with which they can be resolved both change proportionately.

The second advantage of the arrested state derives from enhanced resistance to various *external* physical and chemical hazards (Table 10.3). Although we need not reiterate these in detail here, it is worth reminding the reader again of how metabolic arrest can protect the organism against remarkable stresses, whether they be due to extended anoxia, hypothermia, or dehydration. A part of this enhanced resistance to environmental hazard may be a spin-off of the arresting mechanisms used. Cold torpor in carp, for example, leads to a classic segregation of the nucleolar components of hepatocytes (Saez et al., 1984) which is considered to be a fundamental mechanism in *arresting* DNA-dependent RNA synthesis beyond the level expected from a decrease in temperature by itself. In addition to a controlled suppres-

Table 10.3 Advantages of metabolic arrest as a response to periods of environmental stress

Consequence of metabolic arrest	Functional advantages
Effective length of stress period reduced in proportion to degree of metabolic suppression (time extention)	Substrate depletion and self-pollution both minimized
Enhanced stability of cell structure components (genes, enzymes, membranes, organelles)	Enhanced resistance to environmental hazards

sion of function, such segregation of the nucleolar components may well be a means for *stabilizing* these elements, analogous to similar processes involved in the protection of enzymes, mRNA, and membranes in anhydrobiotes, and thus may account for enhanced resistance to environmental stresses as a result.

Anhydrobiotic organisms, as may be anticipated (Chapter 9), have carried these adaptational mechanisms to their limit and display the most developed resistance to extremes in the physical environment of any animals thus far studied. These organisms, when dried and ametabolic, are able to survive temperature oscillations of over 330°C; they are able to survive in a vacuum, and they will tolerate exposure to high doses of X rays, γ rays, fast neutrons, high-energy electrons, proton beams, or ultraviolet radiation. It is probable, although not yet unequivocally established, that this impressive resistance to harsh physical parameters arises from the loss and redistribution of cell water (Chapter 7) and from its "replacement" (at least in some functions) by substitutes such as trehalose, glycerol, and sorbitol, a process greatly stabilizing macromolecular structures such as proteins, membranes, and intracellular organelles (Crowe et al., 1983). Be that as it may, the extreme resilience expressed by dried anhydrobiotes, matched only by dormancy states in microorganisms (Sussman and Halvorson, 1966), serves to underline a most significant advantage of metabolic arrest as a strategy of defense against environmental hazards.

Metabolic Arrest and the Origin of Life

The paradox is that metabolic arrest in the extreme examples of the anhydrobiotes may be *too* successful as a mechanism of defense against environmental extremes; that is, the enhanced resistance to physical and chemical hazards seems in excess of need, for nowhere on this planet will temperatures near 0°K be found; nowhere on this planet will megarad doses of ionizing radiation be encountered; nowhere on this planet will an organism find itself living in a vacuum. These conditions, however, are common in outer space, and it has been empirically demonstrated that anhydrobiotes such as dried *Artemia* cysts housed on the outside of spacecraft are able to survive exposure to space (Chapter 9). Metabolically arrested bacterial spores or cysts undoubtedly have comparably enhanced resistance to conditions that are common in space, but not on earth (Hanson, 1976).

Such observations have compelled several scientists to consider theories of the extraterrestrial origin of life on earth. The first closely reasoned theory of this sort appeared in a remarkable and interesting publication by Svante Arrhenius in 1908. Although it was logically consistent and contradicted no known scientific laws or observations of the time, the theory got nowhere, mainly because it was not testable. It led to no useful further measurements or experiments, and for that reason lay dormant for some 70 years. Instead of attempting to test and refine the Arrhenius theory, most biologists and biochemists of this time who were interested in the origin of life were inspired by the ideas of Oparin (1924) and Haldane (1929). These ideas led to obvious laboratory experiments, to some success at defining conditions under which inorganic molecules could "evolve" into organic ones (possibly even into biomolecules), and to the widespread acceptance of a chemical evolution theory of the origin of life on this planet (Fox and Dose, 1977).

Despite being almost universally accepted, there have always been problems with the chemical evolution theory. Although examining this theory is beyond the scope of our book, it may be useful to point out that one of its most debilitating difficulties is with time (or rather with the lack of enough time), a difficulty greatly exacerbated by the discovery of evidence of life at microbial cell-level organization in the very oldest sedimentary rocks on the planet. The problem arises because these oldest microfossils on earth, dated as 3.8 billion years old, are only fractionally younger than the earth itself (4.5 billion years). What is worse is a similar time gap of about 0.6 billion years between the origin of the moon and the ages of the oldest lunar rock formations, a finding suggesting that until about 3.9 billion years ago the surfaces of the earth and the moon were highly disturbed and unstable, most probably inhospitable to life or even to biomolecules. Thus the best available data suggest that life appeared on the earth very soon after conditions became favorable for its survival. Chemical evolution had about 0.7 billion years, at the most, to shape inorganic molecules into cells; at the least, there may have been less than 0.1 billion years. The possibility of so massive a step, of going from simple inorganic molecules to cells, in so short a time period is considered by some scientists to be too implausible to consider seriously. That may be why several recent attempts have been made to refine the Arrhenius panspermia theory that requires propagation of living material through space (see, for example, Robinson, 1966, 1967; Crick and Orgel, 1973; Hoyle and Wickramasinghe, 1981; Crick, 1981). We cannot take the time here to seriously assess the current status of these novel hypotheses; suffice it to mention that central to them all is the assumption that any organisms potentially transferable through space must necessarily display greatly enhanced resistance to harsh conditions and must be capable of sustaining fully arrested or ametabolic states for very long time periods indeed. Although there is basically no quantitative way of estimating how long fully dried anhydrobiotes could survive under conditions typically encountered in space, it is widely assumed by workers in this area that only microorganisms, of all known living things, could manage the journeys demanded by most theories of extraterrestrial origins of life (Hanson, 1976). Because at the subcellular level, anhydrobiotic and freeze-tolerant organisms probably utilize defense measures that are similar to those of microorganisms under similar conditions, they are obvious animal candidates for any future exploratory space probes with living material from earth. In a sense, then, our analysis in this book is consistent with the assumptions of the above theories and perhaps lends them added plausibility. But whether or not one of these cosmic theories turns out to be a valid description of the origin of life on earth of course remains an open question.

A Continuum of Metabolism and of Time Extension

A final question of interest is whether metabolic arrest capacities are specialized oddities expressed by only a few groups of unusual animals, each struggling up its own evolutionary dead-end trail. Or are we dealing with a general, and widely distributed, strategy of adaptation with a small number of universally distributed arresting mechanisms? Obviously at this stage we cannot be certain. Nevertheless, it is our impression (perhaps from the way in which we organized the available data) that organisms can be arranged along a temporal or a metabolic continuum. At one end of the continuum are organisms that, in metabolic terms, are unable to suppress metabolism below SMR because so-called maintenance functions must be continuously primed by ATP. From this extreme, where SMR is always SMR, we can work our way through systems that are more and more capable of turning down metabolism below SMR, ultimately to the ametabolic limit. In temporal terms, the same continuum represents organisms with ever-increasing capacities to extend biological time, to the limit in ametabolic systems where time, for practical purposes, has stopped and the time extension factor approaches an infinite value. The existence of such a continuum and the occurrence of metabolic arrest capacities in essentially all major phylogenetic groups that have been analyzed suggest to us that we are dealing with a truly universal strategy of defense against environmental hazards.

There is another line of reasoning leading us to the same conclusion. One might expect at first glance, for example, that the ability to metabolically arrest may involve a multitude of adaptations, whether exhibited at the cellular or the whole-organism level. Organisms capable of arrest might therefore be equipped through phylogenetic history with such specialized compositions, structures, and metabolisms that it would not be possible to arrest cells or organisms that have not sustained a similar adaptational history. We suspect this is not so. In our analysis of diving animals, for example, we intentionally chose to illustrate metabolic defense strategies (which include arrest mechanisms in hypoperfused tissues), using mammals as models. Marine mammals after all are not, and cannot be, all *that* different from other mammals or man. The same point can be made using numerous examples, but perhaps one of the most dramatic involves the capacity of cell membranes to reversibly dehydrate, a precondition for entering reversible ametabolic states that revolves around the properties of membrane phospholipid bilayers.

Phospholipids of biological membranes in bulk H_2O normally exist as bilayers, with proteins and lipids held in this configuration by hydrophobic interactions. When challenged with dehydration or freezing stress, many phospholipids undergo phase transitions, the best known of these being the transition to the hexagonal II (H_{II}) phase, in which phospholipids form long cylinders, with the polar head groups oriented into an aqueous core. The strongest evidence for the H_{II} phase derives from pure phospholipid systems, using a variety of physical techniques (such as X-ray diffraction, ^{31}P-nuclear magnetic resonance, and freeze fracture), but dehydration-induced H_{II} is also demonstrable in natural biological membranes. Except for membranes undergoing fusion, this phase transition is invariably disruptive to membrane functions, which is presumably the selective pressure leading to defensive measures in anhydrobiotic organisms. Cells from these organisms are capable of escaping membrane transitions during dehydration or freezing stresses, most probably because these membranes are stabilized by polyols. Particularly instructive are recent data obtained by Crowe et al. (1983), which show that membranes from unadapted organisms (those typically incapable of sustaining reversible dehydration) can be fully dehydrated in the presence of trehalose, with no evidence of the phase transitions normally accompanying the process. Crowe et al. propose that this protective effect is due to an association of polyols with phosphate head groups of phospholipids, involving (1) a replacement of water associated with the head groups and (2) a change in packing density of the phospholipids under influence of the carbohydrate.

Although the interpretation should be confirmed further, what is most impressive is that an almost trivial mechanism, both in a metabolic sense (the making of a polyol) and in an experimental sense (demonstrating its protective effect), is adequate to "convert" a desiccation-sensitive cell membrane into a dehydration-tolerant one.

This example strikes us as in no way unusual and is one of many that imply that the number of specialized adaptations separating organisms capable of metabolic arrest strategies from those incapable of activating such mechanisms is smaller than would be anticipated on first glance. In fact, the large number of examples of metabolic arrest capacities expressed in vertebrate and invertebrate organisms may be taken as an empirical demonstration that mechanisms underlying these strategies may be harnessable by literally *any* organism. Transforming a sensitive cell, organ, or organism into one resistant to any given environmental hazard may require relatively few modifications, and it may well be possible ultimately to do so in a controllable way with organisms currently considered to be environmentally sensitive. Lamarck may have had intimations of this idea when he wrote in his *Zoological Philosophy* some century ago that "the rotifer of Spallanzani, which was several times reduced to a state of death by rapid desiccation, and afterwards restored to life by being plunged into water, shows that life can be alternately suspended and renewed: it is therefore only an order and state of things in a body." Our analysis shows that less extreme versions of such cryptobiosis or latent life are widely used as defense strategies against environmental stresses; indeed, the potential for them may be universal.

Some capacity for latent life may exist in all of us; what is more, it may be good for us.

Appendixes / References / Index

APPENDIX A

Interactions between O_2 and Metabolism

Analyses of determinants of SMR or BMR frequently focus on variables such as organ work rates, hormonal status (for example, hypo- or hyperthyroidism), caloric status (starved versus satiated versus obese), and ionic concentration and composition (in the case of aquatic organisms). Any changes in SMR related to these parameters can be largely accounted for by obvious changes in the rates of ATP turnover at the cell and tissue level; such SMR changes are readily understandable as arising from alterations in cellular work rates. Somewhat more perplexing and more important for our theme, however, are observations showing that some tissues and organs of mammals at rest display O_2 consumption rates ($\dot{V}_{O_2}$) that vary with O_2 availability. Typically, the higher the O_2 availability, the higher the $\dot{V}_{O_2}$, which reaches a plateau in some cases but not in others (such as some skeletal muscles). The key to this metabolic response may supply us with a key to controlled SMR and thus to controlled metabolic arrest states (Chapter 2).

O_2 Regulators

At the outset, we should emphasize that not all tissues and organs show this response. The brain is perhaps the best example of the opposite pattern, with the cerebral metabolic rate (CMR_{O_2}) being maintained nearly constant over quite large changes in oxygen availability (Siesjo, 1978). Such tissues, organs, or organisms are termed O_2 regulators. When O_2 availability becomes too low, SMR (or, in the case of the brain, CMR_{O_2}), necessarily falls, but the energetic shortfall is at least partially made up by anaerobic glycolysis, the so-called

Pasteur effect (Appendix B). Activation of the Pasteur effect at the cell level can be viewed as a means for maintaining near-normoxic ATP turnover rates despite reduced O_2 availability. Glucose fermentation to lactate generates 2 mol ATP/mol glucose, compared with 36 mol ATP/mol glucose fully oxidized in aerobic metabolism; so to glycolytically make up the shortfall in oxidative ATP production, about 18 times more glucose must be consumed per unit time per unit mass of tissue. However, such large Pasteur effects are rarely seen; for example, a 14-fold activation of glucose consumption in anoxic sperm is one of the largest Pasteur effects we have been able to find recorded in the literature (Hammerstedt and Lardy, 1983). By comparison, over short (6–10 min) periods of O_2-limited function, the Pasteur effect in the mammalian brain makes up 50% of the O_2 deficit (which would represent about an order-of-magnitude increase in glucose consumption rates under these conditions). During longer periods of brain anoxia, the magnitude of the Pasteur effect declines and makes up for even a smaller fraction of the energetic shortfall (Kinter et al., 1984). Nevertheless, organs and organisms whose O_2 consumption rates are largely independent of O_2 availability over broad ranges (the so-called O_2 regulators), typically display powerful Pasteur effects (glucose consumption rates activated 5- to 15-fold during during O_2 lack), in an attempt to maintain normoxic ATP turnover rates. As we will see, this does not occur in O_2 conformers.

O_2 Conformers

Organs and organisms whose $\dot{V}_{O_2}$ varies directly with O_2 availability are termed O_2 conformers, to contrast them with O_2 regulators. The metabolic response of the mammalian brain to changes in O_2 availability fits the pattern of O_2 regulators and represents an extreme end of a spectrum of responses observable in mammalian tissues. Liver expresses an intermediate pattern, whereas skeletal muscle is perhaps the most O_2 conforming of all mammalian tissues (Grubb and Folk, 1978). In liver, as O_2 delivery (the sum of both hepatic arterial and portal flows and O_2 contents) declines, liver $\dot{V}_{O_2}$ initially remains constant, but even at fairly high O_2 delivery rates, $\dot{V}_{O_2}$ begins to decline (Edelstone et al., 1984; Lutz et al., 1975). This complex O_2 conforming pattern of $\dot{V}_{O_2}$ varying with O_2 availability up to some critical value (Chapter 2) is probably the most common pattern for mammalian cells and tissues. Even at the whole-organism level (at least in

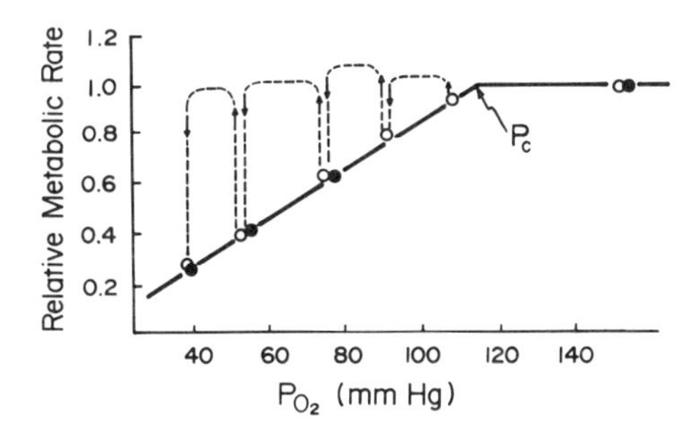

Figure A.1 Hypoxia-dependent metabolic arrest in a small mammal is shown by a decrease in O_2 consumption relative to normoxic rates during successive exposures of 30–40 min at reduced P_{O_2} (closed circles). Because the animals were in negative heat balance below the critical partial pressure (P_c) and because the ensuing hypothermia might influence metabolism independently of hypoxia, a second set of experiments are needed (open circles). Repeated three weeks later with the same animal, but with a 1-hr normoxic recovery after each hypoxic exposure, these tests indicated that the degree of metabolic depression clearly is not compromised by varying body temperature. Modified from Rosenmann and Morrison (1974).

several small mammals), $\dot{V}_{O_2}$ increases as arterial O_2 pressure increases toward a critical (usually species–specific) value, above which $\dot{V}_{O_2}$ is essentially constant no matter how much more O_2 tensions or delivery are increased (Figure A.1). Similar O_2 conforming patterns are common among numerous invertebrate groups and among ectothermic vertebrates. In some invertebrates (for example, lobsters) $\dot{V}_{O_2}$ appears to increase with P_{O_2} essentially indefinitely, but this does not seem to ever be evident in mammals. However, the $\dot{V}_{O_2}$ of skeletal muscle at least in some mammals is known to increase with no apparent limit (Whalen et al., 1973); no plateau in O_2 consumption is reached, even at very high O_2 tensions or very high O_2 delivery rates (Chapter 2).

In all these systems, whose $\dot{V}_{O_2}$ declines as O_2 availability declines, accelerated glucose consumption would be expected to make up the energetic shortfall. However, in contrast to O_2 regulators, a pronounced Pasteur effect does not occur, and anaerobic glycolytic flux may even be depressed (Hochachka, 1982 a,b; Hochachka and Dunn, 1983). In addition to this unexpected glycolytic regulatory response, O_2 conformity at the organ, tissue, and cell levels presents us with another paradox, which revolves around the way mitochondrial metabolism interacts with molecular O_2. This stems from the fact that in all isolated mitochondria thus far studied, the apparent K_m for O_2 is very low, perhaps in the 0.1–0.5 μM range. The paradox is that in most and probably in all these cases the O_2 concentration in arterial plasma exceeds by large factors (10-fold or more) the values that would be required to fully saturate mitochondrial metabolism. Why then should O_2 conforming tissues and organisms behave as if they were truly O_2 limited? The current literaure seems to have three ways to look at this important problem, which we need to briefly analyze in order to appreciate O_2 conformity as a mechanism for suppressing SMR.

O_2 Sensing and O_2 Conformity

One possible explanation for O_2 dependence of $\dot{V}_{O_2}$ assumes a mechanism for sensing O_2 concentration (Mulligan and Lahiri, 1981) either in the tissue or in the plasma; as O_2 concentration declines, this sensing system presumably slows down oxidative metabolism by conventional means; for example, by state 4 to state 3 transitions or by ADP limitation. (Respiratory state 3 of mitochondria is the condition in which all required components are present and the electron transfer system [ETS] is itself the major rate-limiting factor. State 4 is the condition in which only ADP is lacking, often called the controlled or resting state.) In this view, the high O_2 affinity of isolated mitochondria is accepted and assumed applicable to the in vivo situation. Mitochondrial metabolism would in essence be fully saturated with respect to O_2 under most in vivo conditions (over most of the O_2 conforming range). This interpretation is consistent with recent measurements of myoglobin saturation in mammalian skeletal muscle freeze-clamped in situ (Connett et al., 1984). Freezing times in different metabolic states in such studies take only about 5 msec, and such measurements at no point show O_2 limitation to muscle mitochondria (see below for limitations with this approach). The problem with this explanation of O_2 conformity is that metabolically linked O_2 sensors have not been identifiable thus far for any tissues or cells. Wilson and his colleagues have argued that there is no need for some special O_2 sensing mechanism, because the electron transfer system on its own is able to behave as an O_2 sensing device (Nuutinen et al., 1982; Wilson et al., 1979), but this concept has not gained general acceptance. So, the idea that O_2 availability serves as a *set point* for mitochondrial metabolism and regulates oxidative metabolism by conventional signals rather than by true O_2 (substrate) limitation remains a viable but not well established concept.

Control of Oxygen Affinity and O_2 Conformity

A second hypothesis proposed by Wilson and his group encompasses both the in vitro data on high O_2 affinities of isolated mitochondria and the data on living cells and tissues showing much higher apparent K_m values for O_2. According to this model, a decrease in oxygen concentration to a new steady state causes a decrease in the rate of ATP synthesis as a result of a change in the kinetics of cytochrome oxidase. The rate of ATP utilization remains unchanged, however, and a progressive fall in [ATP] and increase in [ADP] and [P_i] occur. This continues until the stimulating effect of decreased [ATP]/[ADP][P_i] on the respiratory rate results in a new steady state in which the rate of ATP synthesis (respiration) is again equal to the rate of ATP utilization. Such a "compensation" is primarily expressed as reduction of cytochrome *c* and activation of cytochrome *c* oxidase activity.

In essence, the in vitro and in vivo data on the oxygen dependence of mitochondrial oxidative phosphorylation are rationalized by this model by assuming that the apparent K_m for oxygen is strongly dependent on the state of reduction of cytochrome *c* and on the [ATP]/[ADP][P_i]. This would allow the in vivo oxygen dependence to differ from that normally observed for mitochondria in vitro. Wilson recognizes that resolution of this question will probably not be complete until the details of the mechanism of oxygen reduction by cytochrome *c* oxidase and its regulation are known. Meanwhile, to accept this hypothesis we must also accept a model of the cytochrome oxidase catalytic cycle that incorporates a reversible O_2-binding step occurring before a final *irreversible* reaction. According to this model, the O_2 dependence of respiratory rate can then be expressed via the reversible step; and, because all preceding reactions to this point in the ETS are reversible, the redox state of the ETS and the phosphate potential can then be readily adjusted according to changing O_2 availability (Wilson et al., 1979).

True O_2 Limitation and O_2 Conformity

A third explanation of the O_2 dependence of $\dot{V}_{O_2}$ assumes true substrate (O_2) limitation to mitochondrial metabolism. This indeed is the simplest explanation for any such results and would undoubtedly be the favored interpretation were we dealing with any substrate other than O_2. For example, a similar glucose titration of brain metabolism is almost automatically assumed to represent CMR_{O_2} moving up or down a glucose saturation curve (either for the transporter facilitating glucose flux into brain cells or for hexokinase catalyzing its entry into cell metabolism per se). The reason this explanation is not the preferred one for O_2 saturation is because isolated mitochondria show such a remarkably low K_m for O_2, and this observation is not disputed by anyone in the field. Studies with cell suspensions, however (as indicated above), often report apparent K_m values that are 10–100 times *higher* than the K_m values for isolated mitochondria (Kennedy and Jones, 1986). Moreover, organs and tissues frequently behave as if their oxidative metabolism displayed a much higher K_m for O_2, and often this low apparent affinity for O_2 also seems to be true for the whole-organism level. Habeler and Messner, when scaling Mt. Everest without supplementary O_2, certainly felt as if their working muscles were *truly* O_2 limited even if the O_2 concentration in arterial plasma was still in the range of 40 to 50 μM, perhaps over a 100 times higher than the K_m for isolated mitochondria. The hypothesis of O_2 diffusional limitation assumes that in vivo mitochondria represent deep O_2 sinks because of their high rates of O_2 utilization and that as a result sharply defined and steep O_2 concentration gradients lead to diffusional limitation to mitochondrial energy metabolism. This diffusional limitation would then necessarily lead to a large elevation in the apparent K_m for O_2 and would bring in line the in vitro and in vivo data.

Although the diffusional limitation hypothesis can explain O_2 conformity, there has been little direct evidence for it. Longmuir and his colleagues (Benson et al., 1980) have tried to assess the distribution of molecular O_2 in living cells using an intracellular, O_2-sensing fluorescent dye and have marshaled evidence for a patchy or heterogeneous O_2 distribution (that is, anoxic versus oxygenated zones). Others, also using a dye capable of sensing when the electron transfer system is functional, find that patches of mitochondria in living cells periodically switch metabolism on and off (consistent with mitochondria in specific cell regions periodically being O_2 depleted). Jones and his coworkers (1982, 1984) have looked at the redox state of electron transfer systems in isolated myocytes and hepatocytes and have shown that the O_2 concentration for half-saturation may exceed the K_m for isolated mitochondria by up to an order of magnitude. All of these data, plus earlier work, reviewed by Chance (1976), are clearly consistent with the idea that diffusional limitations raise the apparent

K_m for O_2 so much that metabolism may be truly O_2 limited. However, they are inconsistent with the recent freeze-clamp data on myoglobin binding of O_2, which implies no significant O_2 gradients in living cells (Connett et al., 1984). Kennedy and Jones (1986) counter these data with direct measurements showing that the oxidation state of cytochrome a_3 of isolated myocytes parallels myoglobin oxygenation, a result that indicates O_2 concentration "sinks" at the inner mitochondrial membrane.

Wilson has used the analogy of an O_2 electrode to argue against the probability of O_2 diffusion limitation in vivo. Oxygen electrodes are available with dimensions similar to those of mitochondria; and, because functionally they behave as total oxygen sinks, the generated oxygen gradients can be accurately assessed. A clean platinum oxygen cathode, for example, develops currents of approximately 2.6×10^{-12} A mm Hg^{-1} μm^{-2} surface area, which may represent a diffusion limit for O_2 in simple aqueous solutions. In contrast, assuming mitochondria to be 1-μm spheres containing 0.5 nmol cytochrome a_3 per milligram of protein, 2 μl volume per milligram of protein, and a respiratory rate of 12 nmol O_2 per nmol cytochrome a_3 per second, O_2 uptake rate can be shown to be equivalent to an electrical current of 1.85×10^{-13} A/μm^2 surface area. Thus, even at an oxygen pressure of only 1 mm Hg, a micro oxygen electrode consumes ten times more oxygen per unit surface area than does a mitochondrion at saturating oxygen concentrations.

If the electrode analogy is valid for intracellular conditions, it is unlikely that there are steep oxygen diffusion gradients to the mitochondria. Unfortunately, it may not be wise to make such extrapolations, for at least two reasons. First, the solubility coefficient for O_2 may well vary in different locations in the cell; evidence along these lines, for example, arises from the work of Longmuir and his colleagues (Benson et al., 1980) using O_2 quenching of fluorescence emission of pyrenebutyric acid to quantify O_2 distribution in living cells. Second, local variations in intracellular viscosity may occur within the cell, and this too would invalidate analogies between O_2 electrode behavior in simple aqueous solution and mitochondrial behavior in living cells. Evidence along these lines arises from studies on intracellular diffusion by Horowitz and Paine (using an ingenious method called the "reference phase technique"); these studies (1980, 1981) show that diffusion rates of large molecules (proteins) are 1/10 to 1/100 those expected in water and that diffusion even of small neutral molecules (for example, sucrose) within cells displays regional variability within cells (Mastro et al., 1984). The reference phase technique is potentially transferable to studies of intracellular diffusion and transport of gases, but to our knowledge this has not yet been achieved. Nevertheless, from estimates of intracellular O_2 diffusion coefficients already available (Jones and Kennedy, 1986), the simple extrapolation of O_2-electrode functional properties in water to mitochondrial properties in vivo would not seem to be prudent.

If only for the sake of adopting a position, then, we conclude that current evidence on balance most favors the hypothesis of diffusional limitations as the explanation of apparent K_m differences in vivo and in vitro and the O_2 conformity that is often observed in vivo, with the proviso that the location of the largest gradient is still an open question (Honig et al., 1984). Perhaps the most convincing evidence to our minds arises from a comparison of the O_2 dependence of different intracellular O_2-requiring functions: myoglobin binding of O_2, reduction of mitochondrial ETS components, and urate oxidase activity. As O_2 availability declines, all three functions decline in parallel despite very different in vitro O_2 dependencies (very different in vitro K_m values). Over a decade ago, Chance (1976) took these results to indicate very steep O_2 concentration gradients in the cell, with O_2 concentration at the mitochondria being much lower than would be anticipated from O_2 tensions in venous effluent. To us, that position still seems the most defensible today.

APPENDIX B

Normal and Reversed Pasteur Effects

The Pasteur effect, named for its discoverer, who worked out the basic phenomenon with yeast in 1861, is defined as the inhibition of carbohydrate (glucose) consumption when O_2 concentrations are high. It also includes the opposite situation: increased anaerobic glycolysis when O_2 is limiting. A reversed Pasteur effect is defined as *decreased or unchanging glycolytic flux when O_2 is limiting;* and it must be carefully distinguished from the Crabtree effect, which is defined as the inhibition of O_2 consumption by activated carbohydrate fermentation. In terms of modern metabolic biochemistry, then, a Pasteur effect is activated glycolysis when O_2 is limiting; a reversed or negative Pasteur effect, is literally and simply the absence of such an activation process. Although underlying mechanisms are still not fully resolved, current approaches seem to fall naturally into three categories: (1) allosteric regulation of glycolysis, (2) covalent modification of key regulatory enzymes, and (3) enzyme and pathway functional compartmentalization. We will look at each of these in turn.

Conventional Metabolite Modulation

Historically the most interesting hypothesis attempting to explain the Pasteur effect emphasized P_i in regulating glycolysis. The idea is that glycolysis and oxidative phosphorylation compete for inorganic phosphate; so, in the presence of oxygen, the latter limits the availability of P_i for glycolysis. This brilliant, but simple idea, which quickly became popular, suffers a serious deficiency: it only explains an inhibition of lactate formation, which is dependent on P_i, but it does not explain the original observation by Pasteur of O_2-dependent inhibition of glucose phosphorylation, which is independent of P_i and requires only ATP, a product of oxidative phosphorylation.

Racker (1976), reviewing how our concepts have developed, points out that when this problem with the P_i-based model was realized, some time went by before evidence for allosteric models began to accumulate and lead to a somewhat more sound concept of coordinated control of glucose metabolism as a function of O_2 availability. The fundamental framework for this approach was initially, and still is, centered on an interplay between key regulatory enzymes in the pathway. Usually the emphasis is on hexokinase (HK) and phosphofructokinase (PFK), for HK is inhibited by glucose 6-phosphate (G6P) and PFK is inhibited by high concentrations of ATP. These two enzymes are inhibited at physiological concentrations of the allosteric effectors, and the two intermediates thereby influence the rate of glucose utilization. The Pasteur effect can thus be formulated as the result of a set of sequential and cascading events initiated as O_2 availability for oxidative phosphorylation increases: (1) depletion of P_i and ADP required for glycolysis and increase in the ATP level, (2) inhibition of phosphofructokinase by ATP and a subsequent G6P accumulation, and (3) inhibition of hexokinase by G6P.

Because of cell specializations, the mechanism of the Pasteur effect may need to be modified in some cases. In yeast, for example, HK is not under G6P product inhibition, so this mechanism cannot contribute to control. It is not surprising, therefore, that ^{31}P and ^{13}C NMR spectroscopic studies in yeast can *fully* account for an observed Pasteur effect (of about a 2-fold increase in glycolytic flux on transition to anoxia) by changes in flux at the PFK locus; these changes are mediated by increased $F2,6P_2$, $F1,6P_2$, ATP, AMP, P_i, and H^+ concentrations (Reibstein et al., 1986). In a different kind of preparation (permeabilized yeast cells metabolizing G6P), increased glycolytic flux is sustained by coupling the pathway to AMP deaminase; the latter is thought to serve in stabilizing adenylate ratios (by adenylate depletion), with the undesirable effect that this takes away at least three potential PFK activators (relatively elevated levels of AMP, ADP, and P_i). In the absence of a secondary backup glycolytic activator, AMP deaminase stabilization of the adenylate ratios might be expected to slow down glycolytic flux, but this does not occur because NH_4^+, which is a well-known allosteric PFK activator, accumulates enough under these conditions to contribute to the sustained glycolytic flux and the observed Pasteur effect (Yoshimo and Murakami, 1983).

One of the more interesting Pasteur effects noted in the literature is that activated by bull sperm during reversible transitions between anaerobiosis and aerobiosis. Because sperm display a simplified metabolism, it is possible to measure both the in situ flux rate at specific enzyme-catalyzed steps as well as the overall net flux and net ATP yield. Such data, analyzed by Regen and Pilkis (1984), can be quantified using the Kacser-Burns (1973) metabolic control theory to specify the contribution of different sites to the Pasteur effect. The principle of such analysis, worked out independently by Kacser and Burns (1973) in the United Kingdom and by Heinrich and Rapaport et al. (1976) in East Germany, is straightforward and involves several theorems showing how the sensitivity of pathway flux to the activity of a component enzyme depends upon coefficients of the pathway. The sensitivity expression they and others have used is the fractional change of pathway rate per fractional change of a specific enzyme activity, an expression termed the "sensitivity coefficient" or more commonly the "control strength" for that enzyme step. Each enzyme in a given pathway has its own characteristic control strength or Z value. If a single enzyme were rate limiting in an absolute sense (flux through this enzyme step being exactly equal to flux of the pathway), the Z value for that enzyme, by definition, would be 1, a situation in fact never observed, as it would imply that other enzymes have Z values of 0. Instead, most enzymes usually contribute *something* to flux controls (usually display finite Z values), and the sum of all Z values is unity (the sum of fractional changes in enzyme activities by definition can never exceed the fractional changes in pathway rate). In the case of bull sperm under normoxia, HK is not very rate determining despite its position at the head of the pathway, because of a large "elasticity" (a large fractional change in the HK net rate per fractional change in metabolite concentrations). As a result, control is passed from the HK step to subsequent steps in the pathways, mainly to PFK, which is the enzyme to which glycolytic flux is most sensitive under O_2-saturating conditions. As glycolysis accelerates by use of specific ETS inhibitors (equivalent to a 14-fold, natural Pasteur effect), PFK control strength is reduced while that of hexokinase is increased to values about three times those for PFK. These subtle changes in the rate limitingness of HK and PFK on transition from low to high glycolytic flux rates are generated by entirely conventional regulatory mechanisms: with G6P deinhibition at the HK step and with adenylate (mainly AMP) control at the PFK step.

To the degree that the preceding analysis of the Pasteur effect is empirical, it is valid irrespective of other controlling inputs. Nevertheless, the implicit assumption in all similar analyses (whether they be made for yeast, sperm, or whole organs) is that changes in flux of the pathway are *caused* by specific modulator-mediated changes in catalytic rate at specific, identifiable enzyme steps in the pathway. However, this need not be the case, for activity changes of enzymes in pathways may be caused by adjusting the ratio of active to inactive forms of enzymes, a proposal that brings us to an alternative way of looking at the problem.

Covalent Modification of Enzymes

One of the most common ways known for modifying the active:inactive ratios of enzymes involves controlled (that is, enzyme-catalyzed) phosphorylation and dephosphorylation cascades. The best worked out example in mammalian tissue involves glycogen phosphorylase, which occurs in a dimeric, low-activity *b* form which, in the presence of *b* kinase, Ca^{++}, and ATP, can be rapidly phosphorylated to a more active, tetramer *a* form. It is obvious that in tissues that store glycogen the equivalent of a Pasteur effect would, in the absence of O_2, lead to increased consumption of glycogen, and not necessarily of glucose. As seen in Chapter 2, hypoxia-sensitive tissues and organs (which typically display reasonably marked Pasteur effects) typically sustain significant Ca^{++} accumulation, which undoubtedly could contribute to *b* kinase activation, followed by phosphorylase $b \rightarrow a$ conversion, and to a net increase in glycolytic flux.

Three-Dimensional Control of Metabolism

A third approach to unraveling the mechanisms of the Pasteur effect places particular emphasis on numerous bits and pieces of evidence that show that the enzymes of glycolysis are not really functional in a so-called cytosol or soluble phase, subject to simple, mass-action dominated control. Instead, these data support the concept of cell metabolism as a highly organized matrix of pathways, where *physical position* (enzymes bound-to-cell ultrastructure or enzymes interacting specifically with other enzymes) is a critical—perhaps the most critical—determinant of activity. Hypoxia and ischemia in mammalian heart, for example, lead to an increased binding of glycolytic enzymes, a process leading to increased enzyme activity and increased glycolytic flux and potentially contributing to a positive Pasteur effect (Clarke et al., 1984). Whereas in this kind of conceptualization (that is, a three-dimensional theory of metabolic control) metabolite concentration changes are still certainly predicted and are still obvious sources of information about flux through the pathway, they are effects, not causes, of change in enzyme activity. This conceptualization, in other words, turns things around completely. Support for these kinds of control concepts is of two forms: on the one hand, numerous workers have obtained evidence for a significant degree of enzyme localization and enzyme order in cell metabolism, which we refer to as structural evidence. On the other hand, more recent studies have marshaled evidence showing that even in solution, enzyme–enzyme interactions are potent determinants of pathway function and this line of support we shall refer to as functional evidence for an important three-dimensional component to metabolic control.

Structural Evidence for Three-Dimensional Metabolic Control Concept

A rather unexpected source of information about how many enzymes of the total complement exist free in solution in eucaryotic cells comes from studies analyzing the forms and behavior of intracellular water (Clegg, 1984). An impressive and rather overlooked early example is the work of Kempner and Miller (1968), in which they stratified *Euglena* (a unicellular eucaryote) by ultracentrifugation (100,000 *g* for an hour). Cytochemistry of macromolecules and enzyme-specific cytochemistry show that no macromolecules of any kind, including a variety of classic "soluble" enzymes, occur at detectable levels in the aqueous band; the cells, which in this treatment serve as their own centrifuge tube, retain viability after such treatment, a result indicating that the outcome may be taken as an indication of events as they exist in the cell. Analogous work on stratified *Neurospora* and *Artemia* cells also indicates that intact cells contain very few truly soluble enzymes or macromolecules; that is, the equilibrium

$$\text{soluble enzymes} \rightleftharpoons \text{bound enzymes}$$

is shifted far to the right under these conditions.

In a totally different approach, studies of intracellular diffusion of endogenous and injected materials (using a reference phase technique) indicate that most intracellular enzymes and proteins do not diffuse at all over time courses of up to several hours, and this appears true in both amphibian eggs and mammalian fibroblasts. More recently, electron spin resonance measurements show that the microviscosity of normal cells and cells 50% dehydrated behave as if the cytoplasm contained only minor amounts of dissolved macromolecules (or none at all). In this context, Clegg and his coworkers find that when L-cells in culture are placed in 2 *M* sorbitol, they may lose up to 80% of intracellular water with hardly a measurable effect on glucose metabolism, an utterly implausible result if the pathway of glycolysis were operative in a simple aqueous solution (Chapter 9).

Most workers in this field consider that the explanation for the data described above is that water in cells exists in various forms (Chapters 7 and 9). Of these, bulk water serves more as a communication channel than it does as a solvent for "soluble" glycolytic enzymes and other so-called cytosolic pathways.

But if glycolytic enzymes are largely out of solution, where are they to be found? The answers here are many and depend in part on the cell or tissue type under analysis (see Clegg [1984] for an in-depth review of this area). Trypanosomes, for example, contain an actual membrane-bound microbody or organelle, termed the glycosome, which houses most of the enzymes of glycolysis (Cannata and Cazzulo, 1984). In skeletal and cardiac muscle, myofibrils have long been considered to possess specific binding regions for glycolytic enzymes. Red blood cells, which in mammals are powered entirely by glycolysis, bind several glycolytic enzymes specifically to band 3 protein. In addition, Clegg has marshaled evidence in support of a binding role for the microtrabecular lattice, an enormously branched series of protein-rich structures that ramify throughout the cell. For all these workers the implications are the same: pathways such as glycolysis, classically considered to operate in simple cytosolic solution, may indeed be structurally compartmentalized. One possible (and extreme) functional consequence, suggested as an intentionally provocative working hypothesis by Ureta (1978), is that different isozyme complexes are structurally strung together, each forming its own semi-independent, functionally discrete pathway. This arrangement could be the basis for the apparent simultaneous coexistence in arterial smooth muscle of one pathway for glucose fermentation to lactate (in the presence of O_2), which is coupled to Na^+,K^+-ATPase function, and another for the fermentation of glycogen to lactate (activated under O_2-limiting conditions), which supplies energy for muscle work. In essence the studies of Lynch and Paul (1983) imply that within a single smooth muscle cell two pathways of glycolysis are structurally and functionally discrete enough so that glucosyl units derived from endogenous breakdown of glycogen do not enter the same pool of glycolytic intermediates that are used in the aerobic production of lactate from exogenous glucose, and vice versa. The picture emerging—two lactate-forming pathways, semi-independent, with discrete starting substrates, and nonmixing pools of chemically identical intermediates, one showing a Pasteur effect, the other a reversed Pasteur effect—is a far cry from the model of glycolysis and similar enzyme pathways that we have in mind when we look at crossover plots, at control strengths, or at elasticity!

Do any of the current three-dimensional metabolic control theories adequately account for the Pasteur effect when it, or its reverse (Storey, 1985), is observed in ectothermic anaerobes? In one sense, it is evident that in principle the *only condition* needed to cause a Pasteur effect is one shifting the equilibrium

soluble glycolytic enzymes $\rightleftharpoons$ bound glycolytic enzymes
(less active forms) (more active forms)

to the right whenever O_2-limiting conditions prevail. A reversed Pasteur effect then could be caused by shifting that equilibrium to the left under hypoxic conditions. In ischemic or hypoxic mammalian heart, as mentioned above, a Pasteur effect is observed and the equilibrium

solubilizable glycolytic enzymes (less active) $\rightleftharpoons$ myofibrillar-bound glycolytic enzymes (more active)

is shifted to the right (Clarke et al., 1984). In contrast, in anoxic bivalve mollusks, the equilibrium is shifted to the left, consistent with the reversed Pasteur effect that is the hallmark of such organisms (Storey, 1985). These are promising beginnings, yet we should emphasize that thus far no single study has been able to show that these effects could be large enough to fully account for the magnitude of the observed Pasteur effect, nor actually do any of these studies show us why bound enzymes (for example, myofibrillar-complexed aldolases) should be better catalysts than the same enzymes are in simple solution. This kind of information necessarily comes from functional studies.

Functional Evidence for Three-Dimensional Metabolic Control Concept

Perhaps the most direct reason for rethinking the way metabolic pathways such as glycolysis are organized in vivo comes from the relative concentrations of proteins and metabolites in cells. Despite a great deal of cell-to-cell variability in the concentrations of specific proteins, the total concentration (in milligrams per milliliter or in molarity of peptide bonds) is remarkably uniform. Estimated localized concentrations of 200–300 mg/ml (Table B.1) are only modestly lower than the concentration in protein crystals used in X-ray crystallography, which indeed is consistent with the high vicosity of all cell cytoplasms.

In view of such high protein concentrations, it is fair to ask, How much room remains for the metabolites of the cell? And how are substrate metabolites transferred from enzyme site to site in such a highly viscous medium? The answer to the first question is, Not much, which is why concentrations must be low; to the second, By direct enzyme-to-enzyme transfer, at least in some cases. These answers imply that structural interactions within the cell are the framework of metabolism. This model can be nicely illustrated by maintaining our focus on glycolysis. Rough estimates of cellular concentrations of individual glycolytic enzymes of muscle sarcoplasm (Table B.2) indicate that two glycolytic components, glyceraldehyde-3-phosphate dehydrogenase, by themselves, represent 40–50% of the total glycolytic enzymes, or over 10% of the protein content of typical protein crystals. In contrast, the concentrations of intermediary metabolites of glycolysis are much lower on a molar basis (Table B.3). Enzyme concentrations at the same level as, or even higher than, substrate concentrations is just the reverse of conditions for most in vitro studies of enzyme catalysis, and this reverse situation places the problem of transfer of metabolite from one enzyme site to another in a

Table B.1 Concentrations of total soluble proteins in different cell types

Source	Protein concentrations (mg/ml cell water)
Red cells (human)	250–320
E. coli	ca. 200
Brewer's yeast	ca. 114
White muscle (rabbit)	220–275
Red muscle (rat)	192–240
Heart (rat)	200–250

Source: From Srivastava and Bernhard (1986).

Table B.2 Concentrations of individual glycolytic enzymes in rabbit muscle sarcoplasm

Enzyme	Concentration (mg/ml)	Site concentration (μ*M*)
Phosphoglucomutase	1.98	31.9
Aldolase	30.35	809.3
α-Glycerol-P dehydrogenase	1.78	61.4
Triose-P-isomerase	5.82	223.8
Glyceraldehyde-3-phosphate dehydrogenase	50.7	1398.6
Phosphoglycerate kinase	5.81	133.6
Phosphoglycerate mutase	6.37	235.9
Enolase	22.98	540.7
Pyruvate kinase	9.94	172.9
Lactate dehydrogenase	11.10	296.0

Source: From Srivastava and Bernhard (1986).

new light. The total concentrations of NAD^+ and NADH in particular are less than the total concentration of dehydrogenase binding sites in the muscle and liver cytosol, NADH occurring at only a small fraction of the total dehydrogenase binding sites. With most of the intermediates in glycolysis being enzyme bound (the clear-cut exceptions being starting fuels [hexose phosphates and phosphocreatine] and end products [lactate and possibly ATP]), the implication is that excessive enzyme site concentrations rather than excessive metabolite concentrations control substrate distribution. One can therefore appreciate the strength of selective pressures favoring some order to enzyme function in this kind of metabolic field. For example, it is difficult to see how the transfer of bound NADH among four cytosolic dehydrogenases—glyceraldehyde-3-P dehydrogenase (GPDH), α-glycerophosphate dehydrogenase (α-GPDH), lactate dehydrogenase (LDH), and malate dehydrogenase (MDH)—could be achieved without some sort of special mechanism. Studies by Bernhard and Srivastava suggest that the special mechanism involves direct transfer of NADH enzyme to enzyme (rather than a dissociation of NADH into solution, followed by an enzyme competition for limiting coenzyme). Their concept is of such significance to understanding metabolic control in general, as well as the Pasteur effect specifically, that we must examine it more closely.

Table B.3 Concentrations of precursors, coenzymes, and glycolytic intermediates

Metabolite	Concentration (μM)[a]
Glucose-1-P	240
Glucose-6-P	3,900
Fructose-6-P	1,500
Fructose-1,6-P_2	80
Dihydroxyacetone-P	68–160
Glyceraldehyde-3-P	80
3-Phosphoglycerol-GPDH	800
1,3-Diphosphoglycerate	50
3-Phosphoglycerate	152–200
2-Phosphoglycerate	20
Phosphenolpyruvate	65
Pyruvate	380
Lactate	3,700
Creatine-P	26,600
ATP	8,050
ADP	926
AMP	43
P_i	8,000
NAD^+	541
NADH	50

Source: From Srivastava and Bernhard (1986).
a. In resting rat muscle, except for the concentrations of coenzymes, which are for liver. Values are based on the assumption that these intermediates are restricted to the cytoplasmic fluid.

The Handoff versus the Fumble: Two Ways to Transfer Metabolites from Enzyme to Enzyme

In principle, metabolites can be transferred from their site of synthesis to their site of utilization by one of two mechanisms, either by dissociation from their site of synthesis into solution, then by diffusion (Equation 1), or by direct transfer from site of synthesis to site of utilization without the intervention of the aqueous solvent (Equation 2):

$$E_1\text{–}S_2 \rightarrow E_1 + S_2 + E_2 \rightarrow E_2\text{–}S_2 \rightarrow \text{products} \quad (1)$$
(random diffusion)

$$E_1\text{–}S_2 + E_2 \rightarrow E_1\text{–}S_2\text{–}E_2 \rightarrow \text{products} \quad (2)$$
(direct transfer)

Srivastava and Bernhard (1986) refer to the latter as a handoff, the former as a fumble; for at least two kinds of enzyme couples in glycolysis (PGK transfer of 1,3-diphosphoglycerate [DPG] to GPDH and GPDH transfer of NADH to LDH), kinetic evidence now strongly favors the handoff alternative. Experiments designed to test this all involve the reduction of metabolite concentration in solution by the formation of enzyme–metabolite complex in the presence of an excess of enzyme over metabolite. We will illustrate the case with PGK. When the kinase is in excess of the substrate, DPG, virtually all of the metabolite is bound to the enzyme, since the K_d for the DPG–enzyme complex is smaller than 10^{-8} *M*; as a result, the aqueous DPG concentration is exceedingly low, and the question arises as to which is a better substrate for GPDH, the PGK–DPG complex or free DPG. In fact, can the reaction

$$\text{PGK–DPG} + \text{NADH} \xrightarrow{\text{GPDH}} \text{PGK} + \text{G3P} + \text{NAD}^+ + \text{P}_i$$

even proceed? In tests designed to answer these questions, the concentration of GPDH utilized is low, comparable to the usual concentrations of enzyme catalyst employed in enzymological assays (of the order of nanomolar concentrations). Any aqueous DPG formed by dissociation of the PGK–DPG complex will almost invariably be reabsorbed by PGK rather than by GPDH. The rate of reaction of the metabolite should thus be substantially slowed down if aqueous DPG is required for the GPDH-catalyzed reaction (because of its low buffered concentration or its very slow rate of dissociation from PGK). This does not occur, however, and the actual rate for the reaction is far higher than can be accounted for either by rates of PGK–DPG dissociation or by rates of DPG diffusion. These and other results are therefore consistent with the handoff or direct transfer mechanism of metabolite transfer between two adjacent enzymes in the pathway.

An interesting feature of the coupled-transfer reaction between PGK–DPG and GPDH is that, at PGK–DPG saturation, the rate can be further enhanced almost 20-fold by the presence of saturating 3-phosphoglycerate (3PGA). At effective 3PGA concentrations, this metabolite is neither an inhibitor of the GPDH-catalyzed reaction, nor

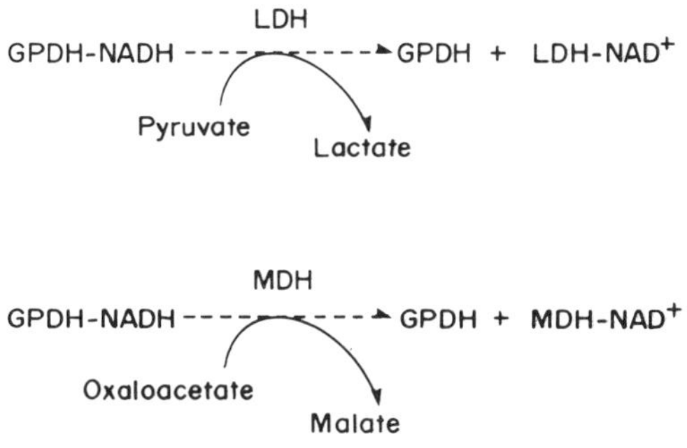

LDH

GPDH-NADH ----------→ GPDH + LDH-NAD^+

Pyruvate → Lactate

MDH

GPDH-NADH ----------→ GPDH + MDH-NAD^+

Oxaloacetate → Malate

Scheme B.1

is it an effective competitive inhibitor of the interaction of DPG with phosphoglycerate kinase. This means that 3PGA is a specific effector of the transfer process itself. The activation is achieved because DPG can be transferred within the enzyme–enzyme complex either toward the GPDH-catalyzed reaction, or back to the kinase site. However, if the vacant kinase site is already occupied by 3PGA, the ability of DPG to transfer back to the kinase will be inhibited whereas the reduction reaction at the GPDH site can proceed unaltered. The net effect is a powerful activation of the overall reaction rate.

Analogous data on handoff versus fumble mechanisms for transfer of NADH between cytosolic dehydrogenases show that the following kinds of reactions proceed at rates substantially faster than they would if the coenzyme were supplied from solution (Scheme B.1). In the context of this chapter, the NADH transfers that are of major interest are (1) GPDH → LDH, (2) GPDH → MDH, (3) LDH → GPDH, (4) LDH → α-GPDH, and (5) LDH → MDH.

On the basis of these results, it appears that whether or not NADH transfer between any two dehydrogenases is possible depends upon whether the two are of the "A" or "B" type. A-type dehydrogenases catalyze the exchange of hydrogen from the A face of nicotinamide, at the C_4 position; B-type dehydrogenases express the opposite stereospecificity. NADH transfer is not possible between two dehydrogenases of the same stereospecificity for a particular C_4 hydrogen, but enzymes of opposite chiral specificity can hand off NADH in either direction (A → B or B → A).

Whereas the functional advantages of these transfer mechanisms are instantly obvious, the problem of how they are possible is not so clearly evident. Information on the quaternary structure of dehydrogenases (LDH, for example) suggests that the binding of NADH results in the closure of the enzyme binding-site cleft; in this holoenzyme conformation there is no apparent way in which bound coenzyme can escape to the surrounding aqueous environment. On the other hand, there is sufficient space for coenzyme to enter and to leave the active site when the enzyme protein is in its unliganded (or apo) form. Presumably, similar holoenzyme → apoenzyme transitions allow for the free passage of NADH out of, and into, the binding sites of all dehydrogenases. When the surfaces around the active sites are juxtaposed for a pair of A-type enzymes, the conformations of the bound NADHs are the same; for NADH to transfer from site to site, it would have to effect a molecular rotation of 180°, which is impossible. That is presumably why the transfer of NADH between dehydrogenases of the same chirality necessarily involves dissociation from the first site into the aqueous medium, followed by diffusion to the second site. In contrast, when the dehydrogenases of opposite chirality are juxtaposed, the nicotinamide ring of one dehydrogenase site has a mirror-image relationship to the nicotinamide binding site of the other dehydrogenase. The coenzyme molecule bound to one site can transfer its nicotinamide to a second structurally complementary site by translation alone without rotation and without consequent dissociation into the aqueous medium. Thus dehydrogenase-mediated direct transfer of NADH can be brought into line with current understanding of dehydrogenase structure, and we are now in a position to see how these properties may come into play in the Pasteur effect.

Pasteur Effect: Outcome of GPDH Handoffs to PGK and LDH

From the preceding considerations it is evident that a Pasteur effect can be simply explained as a transition from solution- or fumble-dominated glycolysis during normoxia to handoff-dominated glycolysis when O_2 is limiting, a process leading to an 8- to 10-fold increase in flux, assuming no other controlling input. During such activated glycolysis, GPDH must be involved minimally in a dual coupling for the effective transfer of its two products (DPG and NADH) to two target enzyme sites, PGK and LDH, respectively. As summarized in Figure B.1, this dual coupling in essence initiates a series of glycolytic cycles, with LDH and PGK both entering and leaving at prescribed steps in the process. Pyruvate is reduced to lactate while LDH is cycling

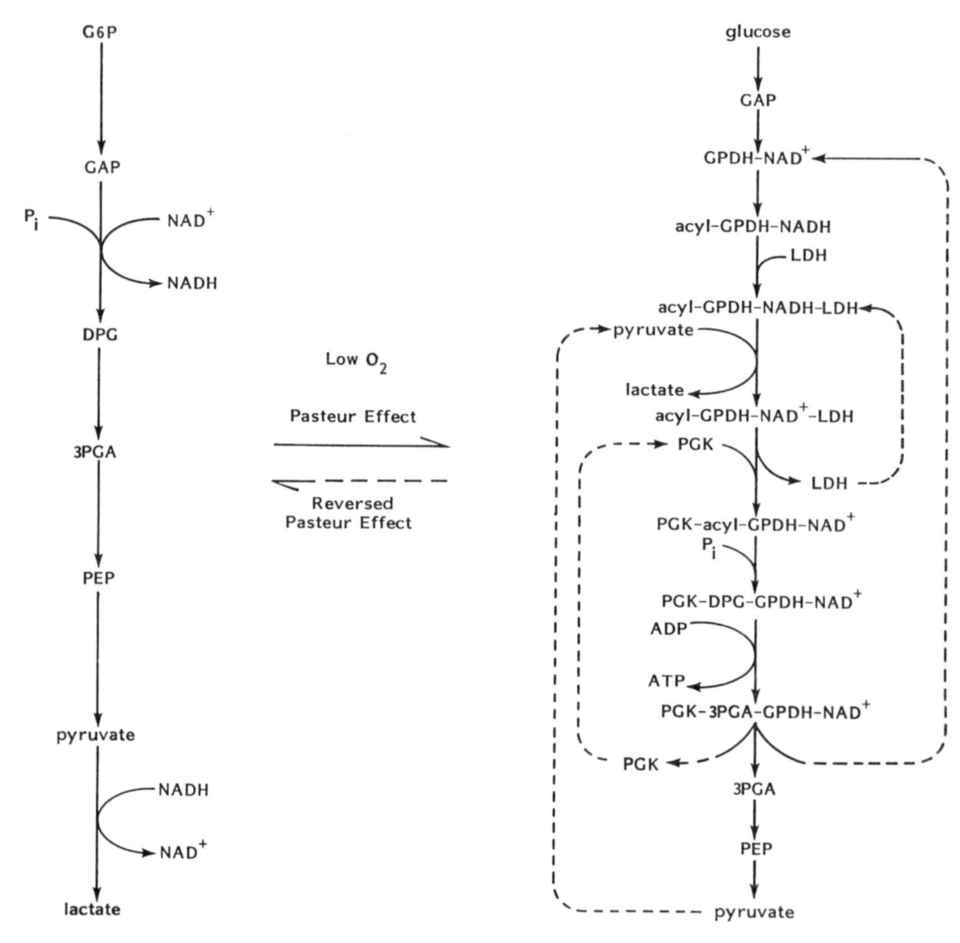

Figure B.1 The Pasteur effect as a transition from solution-dominated glycolytic function to hand-off dominated glycolytic function. Modified from Srivastava and Bernhard (1986).

through; phosphorolysis of the acyl GPDH and subsequent ATP synthesis occur while PGK is cycling through; NADH and NAD^+ never need to enter the aqueous medium, for GPDH–NAD^+, which initiates the entire sequence, is regenerated as 3PGA is released for the next step in glycolysis.

To our knowledge this is the most recent and probably most elegant model of how the Pasteur effect might be mediated, yet it should be emphasized that it has not been proved in vivo. As with the model based on ratios of bound to soluble glycolytic enzymes, no information is available on how large a Pasteur effect could be generated by this system. Nor can this interpretation—at this stage, at least—account for why some cells even within a single organ such as the kidney show very large Pasteur effects (medullary thick ascending limb), some show intermediate ones (proximal tubule cells), and others (distal tubules), only modest ones (Bagnasco et al., 1985). Furthermore, there is no easy way to see how this system might differ in facultative anaerobes, which show a reversed Pasteur effect, nor why GPDH, PGK, and LDH do not usually display very high Z values when the pathway is analyzed by the Kacser and Burns model. Thus, a lot remains to be done; in particular, this functional line of research and binding studies must be coordinated. How, for example, might the binding of GPDH, PGK, and LDH to myofibrillar sites (in muscle) or to band 3 protein (in red blood cells) influence handoff mechanisms. To this date, these two lines of research have been progressing in parallel, and it is our hope they will come together in the near future.

Pasteur Effect May Be Cell-Line and Species Specific

As indicated in Chapter 2, it appears that the most effective defense strategy against anoxia used by good animal anaerobes involves glycolytic arrest of variable degree. It is not yet clear what the main mechanisms utilized may be, but their net effect is expressed as a reversed Pasteur effect. Although the standard Pasteur effect appears to be mediated by any one of three general mechanisms, at this time, a conservative interpretation of available data assumes that there is no single mechanism accounting for it in all tissues and organs. Both its magnitude and its mechanism appear to show a great deal of cell-line specificity arising from cell specializations and adaptations for specific functions. In the case of smooth muscle of arterial walls, the Pasteur effect may even be pathway-specific: one pathway (glucose → lactate) being fully operational in the presence of O_2 and a second (glycogen → lactate) being normally inhibited by high O_2 tensions (Lynch and Paul, 1983).

With this degree of cell-line and species specificity of the standard Pasteur effect, it is reasonable to expect a similar degree of specificity for the reversed Pasteur effect and thus for metabolic arrest in animal anaerobes in hypoxia.

References

1. The Time Extension Factor

Brett, J. R., and T. D. D. Groves. 1979. Physiological energetics. In *Fish Physiology,* vol. 7, ed. W. A. Hoar, D. J. Randall, and J. R. Brett. New York: Academic Press, pp. 279–352.

Calder, W. A. 1981. Scaling of physiological processes in homeothermic animals. *Annu. Rev. Physiol.* 43:301–322.

Hulbert, A. J., and P. L. Else. 1981. Comparison of the "mammal machine" and the "reptile machine": energy use and thyroid activity. *Am. J. Physiol.* 241:R350–R356.

Lindstedt, S. L., and W. A. Calder. 1981. Body size, physiological time, and longevity of homeothermic animals. *Q. Rev. Biol.* 56:1–16.

McGilvery, R. W. 1979. *Biochemistry: A Functional Approach,* Philadelphia: W. B. Saunders.

Rapoport, S. M. 1985. Mechanisms of the maturation of the reticulocyte. In *Circulation, Respiration, and Metabolism,* ed. R. Gilles. Berlin: Springer-Verlag, pp. 333–342.

Schmidt-Nielson, K. 1984. *Scaling—Why Is Animal Size So Important?* Cambridge: Cambridge University Press.

2. Animal Anaerobes

Aprille, J. R., and W. A. Brennan, Jr. 1985. Subcellular distribution of adenine nucleotides as a regulator of metabolic ATP requirements in hypoxic–normoxic transitions. *Mol. Physiol.* 8:648–649.

Austin, J., and J. R. Aprille. 1984. Carboxyatractyloside-insensitive influx and efflux of adenine nucleotides in rat liver mitochondria. *J. Biol. Chem.* 259:154–160.

Berridge, M. J., and R. F. Irvine. 1984. Inositol triphosphate, a novel second messenger in cellular signal transduction. *Nature* 312:315–321.

Brennan, W. A., Jr., and J. R. Aprille. 1985. Regulation of hepatic gluconeogenesis in newborn rabbit: controlling factors in presuckling period. *Am. J. Physiol.* 249:E498–E505.

Brezis, M., S. Rosen, K. Spokes, P. Silva, and F. H. Epstein. 1984. Transport-dependent anoxic cell injury in the isolated perfused rate kidney. *Am. J. Pathol.* 116:327–341.

Catterall, W. A. 1984. The molecular basis of neuronal excitability. *Science* 223:653–661.

Chance, B. 1976. Pyridine nucleotide as an indicator of the O_2 requirements for energy-linked functions of mitochondria. *Circ. Res.* 38:131–138.

Chance, B., and J. S. Leigh, Jr. 1985. The effect of work load upon the operating point for metabolic control in normoxia and hypoxia: "State 4 is best." *News Metabol. Res.* 1(4):26–33.

Edelstone, D. I., M. E. Paulone, and I. R. Holzman. 1984. Hepatic oxygenation during arterial hypoxemia in neonatal lambs. *Am. J. Obstet. Gynecol.* 150:513–518.

Else, P. L. 1984. Studies in the evolution of endothermy: mammals from reptiles. Ph.D. diss., University of Wollongong, N.S.W., Australia.

Farber, J. L., K. R. Chien, and S. Mittnacht, Jr. 1981. The pathogenesis of irreversible cell injury in ischemia. *Am. J. Pathol.* 102:271–281.

Fleckenstein, A., M. Frey, and G. Fleckenstein-Grun. 1983. Consequences of uncontrolled calcium entry and its prevention with calcium antagonists. *Eur. Heart J.* 4(Suppl. H):43–50.

Gatz, R. N., and J. Piiper. 1979. Anaerobic energy metabolism during severe hypoxia in the lungless salamander *Desmognathus fuscus. Respir. Physiol.* 38:377–384.

Hansen, A. J. 1985. Effect of anoxia on ion distribution in the brain. *Physiol. Rev.* 65:101–148.

Hearse, D. J., F. Yamamoto, and J. J. Shattock. 1984. Calcium antagonists and hypothermia: the temperature dependency of the negative inotropic and anti-ischemic properties of verapamil in the isolated rat heart. *Circulation* 70(Suppl. I):154–164.

Hille, B. 1984. *Ionic Channels of Excitable Membranes.* Sunderland, Mass.: Sinauer Associates.

Hochachka, P. W. 1982. Metabolic arrest as a mechanism of protection against hypoxia. In *Protection of Tissues against Hypoxia,* ed. A. Wauquier, M. Borgers, and W. K. Avery. Amsterdam: Elsevier Biomedical, pp. 1–12.

——— 1985. Exercise limitations at high altitude: the metabolic problem and search for its solution. In *Circulation, Respiration, and Metabolism,* ed. R. Gilles. Berlin: Springer-Verlag, pp. 240–249.

——— 1986. Defense strategies against hypoxia and hypothermia. *Science* 231:234–241.

Hochachka, P. W., and J. F. Dunn. 1983. Metabolic arrest: the most effective means of protecting tissues against hypoxia. In *Hypoxia, Exercise, and Altitude. Proc. Third Banff Intl. Hypoxia Symp.* New York: Alan R. Liss, pp. 297–309.

Hochachka, P. W., and T. P. Mommsen. 1983. Protons and anaerobiosis. *Science* 219:1391–1397.

Hochachka, P. W., and G. N. Somero. 1984. *Biochemical Adaptations.* Princeton, N.J.: Princeton University Press.

Hulbert, A. J., and P. L. Else. 1981. Comparison of the "mammal machine" and the "reptile machine": energy use and thyroid activity. *Am. J. Physiol.* 241:R350–R356.

Johansen, K., J. R. Redmond, and C. B. Bourne. 1978. Respiratory exchange and transport of oxygen in *Nautilus pompilius. J. Exp. Zool.* 205:27–36.

Jones, D. P., and F. G. Kennedy. 1982. Intracellular O_2 gradients in cardiac myocytes. Lack of a role for myoglobin in facilitation of intracellular O_2 diffusion. *Biochem. Biophys. Res. Commun.* 105:419–424.

Jones, D. P., F. G. Kennedy, B. S. Andersson, T. Y. Aw, and E. Wilson. 1985. When is a mammalian cell hypoxic? Insights from studies of cells versus mitochondria. *Mol. Physiol.* 8:473–482.

Kinter, D. J. H., J. A. Fitzpatrick, Jr., J. A. Louie, and D. D. Gilboe. 1984. Cerebral O_2 and energy metabolism during and after 30 minutes of moderate hypoxia. *Am. J. Physiol.* 247:E475–E482.

Lykkeboe, G., and K. Johansen. 1982. A cephalopod approach to rethinking about the importance of the Bohr and Haldane effects. *Pacific Science* 36:305–314.

Mangum, C. P., and W. Van Winkle. 1973. Responses of aquatic invertebrates to declining O_2 conditions. *Am. Zool.* 13:529–541.

Matthys, E., Y. Patel, J. Kreisberg, J. H. Stewart, and M. Venkatachalam. 1984. Lipid alterations induced by renal ischemia: pathogenic factor in membrane damage. *Kidney Intl.* 26:153–161.

Nayler, W. G. 1983. Calcium and cell death. *Eur. Heart J.* 4(Suppl. C):33–41.

Palmer, L. G., J. H. Y. Li, B. Lindemann, and I. S. Edelman. 1982. Aldosterone control of the density of sodium channels in toad urinary bladder. *J. Membr. Biol.* 64:91–102.

Sick, T. J., M. Rosenthal, J. C. LaManna, and P. L. Lutz. 1982. Brain K^+ homeostasis, anoxia, and metabolic inhibition in turtles and rats. *Am. J. Physiol.* 243:R281–R288.

Siesjo, B. K. 1981. Cell damage in the brain: a speculative synthesis. *J. Cerebral. Blood Flow Metabol.* 1:155–185.

Spruce, A. E., N. B. Standen, and P. R. Stanfield. 1985. Voltage-dependent ATP-sensitive potassium channels of skeletal muscle membrane. *Nature* 316:736–738.

Stewart, P. A. 1981. *How To Understand Acid-Base,* London: Edward Arnold.

Storey, K. B. 1985. A re-evaluation of the Pasteur effect: new mechanisms in anaerobic metabolism. *Mol. Physiol.* 8:439–461.

Surlykke, A. 1983. Effect of anoxia on the nervous system of a facultative anaerobic invertebrate, *Arenicola marina. Mar. Biol. Lett.* 4:117–126.

Tagawa, K., T. Nishida, F. Watanabe, and M. Koseki. 1985. Mechanism of anoxic damage of mitochondria: depletion of mitochondrial ATP and concomitant release of free Ca^{++}. *Mol. Physiol.* 8:515–524.

Trump, B. F., I. K. Berezesky, and A. R. Osornio-Vargas. 1981. Cell death and the disease process. The role of calcium. In *Cell Death in Biology and Pathology*, ed. I. D. Bowen and R. A. Lockshin. London: Chapman and Hall, pp. 209–242.

Whalen, W. J., D. Buerk, and C. A. Thuning. 1973. Blood flow-limited oxygen consumption in resting cat skeletal muscle. *Am. J. Physiol.* 224:763–768.

3. *Diving Mammals and Birds*

Bert, P. 1870. *Lecons sur la Physiologie Comparee de la Respiration*. Paris: Balliere, pp. 526–553.

Butler, P. J., and D. R. Jones. 1982. The comparative physiology of diving in vertebrates. *Adv. Comp. Physiol. Biochem.* 8:179–364.

Castellini, M. A., B. J. Murphy, M. Fedak, K. Ronald, N. Gofton, and P. W. Hochachka. 1985. Potentially conflicting demands of diving and exercise in seals. *J. Appl. Physiol.* 58:392–399.

Edelstone, D. I., M. E. Paulone, and I. R. Holzman. 1984. Hepatic oxygenation during arterial hypoxemia in neonatal lambs. *Am. J. Obstet. Gynecol.* 150:513–518.

Elsner, R., and B. Gooden. 1983. *Diving and Asphyxia, A Comparative Study of Animals and Man.* Cambridge: Cambridge University Press.

Gotshall, R. W., D. S. Miles, and W. R. Sexson. 1985. Renal oxygen delivery and consumption during progressive hypoxemia in the anesthetized dog. *Proc. Soc. Exp. Biol. Med.* 174:363–367.

Guppy, M., R. D. Hill, R. C. Schneider, J. Qvist, G. C. Liggins, W. M. Zapol, and P. W. Hochachka. 1986.Microcomputer-assisted metabolic studies of voluntary diving of Weddell seals. *Am. J. Physiol.* 250:R175–R187.

Halasz, N. A., R. Elsner, R. S. Garvie, and G. T. Grotke. 1974. Renal recovery from ischemia: a comparative study of harbour seal and dog kidneys. *Am. J. Physiol.* 227:1331–1335.

Hill, R. D. 1986. Microprocessor monitor and blood sampler for freely diving Weddell seals. *J. Appl. Physiol.* 61:1570–1576.

Irving, L. 1939. Respiration in diving mammals. *Physiol. Rev.* 19:112–134.

Issekutz, B., Jr., W. A. S. Shaw, and A. C. Issekutz. 1976. Lactate metabolism in resting and exercising dogs. *J. Appl. Physiol.* 40:312–319.

Kanwisher, J. W., G. Gabrielson, and N. Kanwisher. 1981. Free and forced diving in birds. *Science* 211:717–719.

Kooyman, G. L. 1981. *Weddell Seal, Consummate Diver*. Cambridge: Cambridge University Press, pp. 1–35.

Kooyman, G. L., E. A. Wahrenbrock, M. A. Castellini, R. W. Davis, and E. E. Sinnett. 1980. Aerobic and anaerobic metabolism during voluntary diving in Weddell seals: evidence for preferred pathways from blood chemistry and behaviour. *J. Comp. Physiol.* B 138:335–346.

Lassen, N. A., O. Munck, and J. H. Thaysen. 1961. Oxygen consumption and sodium reabsorption in the kidney. *Acta Physiol. Scand.* 51:371–384.

Lautt, W. W. 1976. Method for increasing hepatic oxygen uptake or other blood-borne substances *in situ*. *J. Appl. Physiol.* 40:269–274.

Lutz, J., H. Henrich, and E. Bauereisen. 1975. Oxygen supply and uptake in the liver and the intestine. *Pflügers Arch.* 360:7–16.

Murdaugh, H. V., B. Schmidt-Nielsen, J. W. Wood, and W. L. Mitchell. 1961. Cessation of renal function during diving in the trained seal (*Phoca vitulina*). *J. Cell. Comp. Physiol.* 58:261–265.

Murphy, B. J., W. M. Zapol, and P. W. Hochachka. 1980. Metabolic activities of heart, lung, and brain during diving and recovery in the Weddell seal. *J. Appl. Physiol.* 48:596–605.

Qvist, J., R. D. Hill, R. C. Schneider, K. J. Falke, G. C. Liggins, M. Guppy, R. L. Elliot, P. W. Hochachka, and W. M. Zapol. 1986. Hemoglobin concentrations and blood gas tensions of free-diving Weddell seals. *J. Appl. Physiol.* 61:1560–1569.

Scholander, P. F. 1940. Experimental investigations in diving mammals and birds. *Hvalradets Skr.* 22:1–131.

Zapol, W. M., G. C. Liggins, R. C. Schneider, J. Qvist, M. T. Snider, R. K. Creasy, and P. W. Hochachka. 1979. Regional blood flow during simulated diving in the conscious Weddell seal. *J. Appl. Physiol.* 47:968–973.

4. Ectothermic Hibernators

Aleksiuk, M. 1976. Reptilian hibernation: evidence of adaptive strategies in *Thamnophis sirtalis parietalis*. *Copeia* 1976, 170–178.

Angelakos, E. T., J. T. Maher, and R. F. Burlington. 1969. Spontaneous cardiac activity at low temperatures in hibernators and non-hibernators: influence of potassium and catecholamines. *Fed. Proc.* 28:1216–1219.

Balaban, R. S., and S. P. Bader. 1984. Studies on the relationship between glycolysis and (Na^+ + K^+)–ATPase in cultured cells. *Biochim. Biophys. Acta* 804:419–426.

Bennet, A. W., and W. R. Dawson. 1976. Metabolism. In *Biology of the Reptilia*, vol. 5, ed. C. Gans and W. R. Dawson. London: Academic Press, pp. 127–223.

Bickler, P. E. 1984. Effects of temperature on acid and base excretion in a lizard *Dipsosaurus dorsalis*. *J. Comp. Physiol.* B 154:97–104.

Carey, C. 1979. Aerobic and anaerobic energy expenditure during rest and activity in *Montane Bufo b. boreus* and *Rana pipiens*. *Oecologia* 39:213–228.

Catterall, W. A. 1984. The molecular basis of neuronal excitability. *Science* 223:653–661.

Cloudsley-Thompson, J. C. 1971. *The Temperature and Water Relations of Reptiles*. London: Merrow Publishing.

Darlington, P. S. 1957. *Zoogeography: the Geographical Distribution of Animals*. New York: John Wiley & Sons.

Derikson, W. K. 1976. Lipid storage and utilization in reptiles. *Am. Zool.* 16:711–723.

Edelman, I. S. 1976. Transition from the poikilotherm to the homeotherm: possible role of sodium transport and thyroid hormone. *Fed. Proc.* 35:2180–2184.

Gatten, R. E. 1978. Aerobic metabolism in snapping turtles, *Chelydra serpentina,* after thermal acclimation. *Comp. Biochem. Physiol.* 61A:325–337.

Glitsch, H. G., and H. Pusch. 1984. On the temperature dependence of the Na pump in sheep Purkinje fibers. *Eur. J. Physiol.* 402:109–116.

Gregory, P. T. 1982. Reptilian hibernation. In *Biology of the Reptilia,* vol. 13, ed. C. Gans and W. R. Dawson. London: Academic Press, pp. 53–154.

Haggag, G., K. A. Raheem, and F. Khalil. 1965. Hibernation in reptiles—I. Changes in blood electrolytes. *Comp. Biochem. Physiol.* 16:457–465.

Hansen, A. J. 1982. Ion and membrane changes in brain anoxia. In *Protection of Tissues against Hypoxia,* ed. A Wauquier, M. Borgers, and W. K. Avery. Amsterdam: Elsevier Biomedical, pp. 199–209.

Hazel, J. R. 1973. The regulation of cellular function by temperature-induced alterations in membrane composition. In *Effects of Temperature on Ectothermic Organisms,* ed. W. Wieser. New York: Springer-Verlag, pp. 55–67.

Hazel, J., and C. L. Prosser. 1970. Interpretation of inverse acclimation to temperature. *Z. Vergl. Physiol.* 67:217–228.

——— 1974. Molecular mechanisms of temperature compensation in poikilotherms. *Physiol. Rev.* 54:620–677.

Heisler, N. 1980. Regulation of the acid-base status in fish. In *Environmental Physiology of Fishes,* ed. M. A. Ali. New York: Plenum, pp. 685–697.

Herbert, C. V., and D. C. Jackson. 1985. Temperature effects on the responses to prolonged submergence in the turtle, *Chrysemys picta bellii:* metabolic rate, blood acid–base and ionic changes, and cardiovascular functions in aerated and anoxic water. *Physiol. Zool.* 58:655–669.

Howell, B. J. and H. Rahn. 1976. Regulation of acid–base balance in reptiles. In *Biology of the Reptilia,* vol. 5, ed. C. Gans and W. R. Dawson. London: Academic Press, pp. 335–363.

Hulbert, A. J., and P. L. Else. 1981. Comparison of the "mammal machine" and the reptile machine: energy use and thyroid activity. *Am. J. Physiol.* 241:R350–R356.

Jackson, D. C. 1978. Respiratory control in air-breathing ectotherms. In *Regulation of Ventilation and Gas Exchange,* ed. D. G. Davies and C. D. Barnes. New York: Academic Press, pp. 93–130.

Jackson, D. C., C. V. Herbert, and G. R. Ultsch. 1984. The comparative physiology of diving in North American freshwater turtles. II. Plasma ion balance during prolonged anoxia. *Physiol. Zool.* 57:632–640.

Johansen, K., and G. Lykkeboe. 1979. Thermal acclimation of aerobic metabolism and O_2–Hb binding in the snake *Vipera beris. J. Comp. Physiol.* 130:293–300.

John-Alder, H. 1984. Seasonal variations in activity, aerobic energetic capacities, and plasma thyroid hormones (T_3 & T_4) in an iguanid lizard. *J. Comp. Physiol.* 154:409–419.

Kutchai, H., and L. M. Geddis. 1984. Regulation of glycolysis in rat aorta. *Am. J. Physiol.* 247:C107–C115.

Li, J. H. Y., L. G. Palmer, I. S. Edelman, and B. Lindemann. 1982. The role of sodium-channel density in the natriferic response of the toad urinary bladder to an antidiuretic hormone. *J. Membr. Biol.* 64:77–89.

Lyman, C. P., J. Willis, A. Malan, and L. Wang. 1982. *Hibernation and Torpor in Mammals and Birds.* New York: Academic Press.

Malan, A. 1983. Adaptation to poikilothermy in endotherms. *J. Therm. Biol.* 8:79–84.

Munday, K. A., and G. F. Blane. 1961. Cold stress of the mammal, bird and reptile. *Comp. Biochem. Physiol.* 2:8–21.

Palmer, L. G., J. H. Y. Li, B. Lindemann, and I. S. Edelman. 1982. Aldosterone control of the density of sodium channels in the toad urinary bladder. *J. Membr. Biol.* 64:91–102.

Pasanen, S., and P. Koskela. 1974. Seasonal and age variation in the metabolism of the common frog *Rana temporaria* L. in northern Finland. *Comp. Biochem. Physiol.* 47A:635–654.

Pollock, M., and E. S. MacAvoy. 1978. Morphological and metabolic changes in muscles of hibernating lizards. *Copeia* 1978, 412–416.

Raheem, K. A. 1980. Taurine—a possible inhibitory substance in the brain of a hibernating reptile. *Comp. Biochem. Physiol.* 65B:751–753.

Raheem, K. A., and W. Hanke. 1980. Changes in the regional distribution of glutamate, aspartate, GABA and alanine in the brain of a lizard *Varanus griseus Daud* during hibernation. *Comp. Biochem. Physiol.* 65B:759–761.

Reeves, R. B. 1977. The interaction of body temperature and acid-base balance in ectothermic vertebrates. *Annu. Rev. Physiol.* 39:559–586.

Rogart, R. 1981. Sodium channels in nerve and muscle membrane. *Annu. Rev. Physiol.* 43:711–725.

Sick, T. J., M. Rosenthall, J. C. LaManna, and P. L. Lutz. 1982. Brain potassium ion homeostasis during anoxia and metabolic inhibition in the turtle and rat. *Am J. Physiol.* 243:R281–R288.

Sick, T. J., E. P. Chasnoff, and M. Rosenthall. 1985. Potassium ion homeostasis and mitochondrial redox status of turtle brain during and after ischemia. *Am. J. Physiol.* 248:R531–R540.

Stevens, E. D. 1973. The evolution of endotherms. *J. Therm. Biol.* 38:597–611.

Stevens, E. D., and M. Kido. 1974. Active sodium transport: a source of metabolic heat during cold adaptation in mammals. *Comp. Biochem. Physiol.* 47A:395–397.

Stoner, H. B., R. A. Little, and K. M. Frayn. 1983. Fat metabolism in elderly patients with severe hypothermia. *Q. J. Exp. Physiol.* 68:701–707.

Ultsch, G. R., C. V. Herbert, and D. C. Jackson. 1984. The comparative physiology of diving in North American freshwater turtles. I. Submergence tolerance, gas exchange and acid–base tolerance. *Physiol. Zool.* 57:620–631.

Walsh, P., and T. W. Moon. 1982. The influence of temperature on extracellular and intracellular pH in the American eel *Anguilla rostrata. Respir. Physiol.* 50:129–140.

Walsh, P. J., G. D. Foster, and T. W. Moon. 1983. The effects of temperature on metabolism of the American eel *Anguilla rostrata (Le Sueur).* Compensation in the summer and torpor in the winter. *Physiol. Zool.* 53:532–540.

Whittam, R., M. E. Ager, and J. S. Wiley. 1964. Control of lactate production by membrane adenosine triphosphatase activity in human erythrocyte. *Nature* 202:1111–1112.

Willis, J. S. 1979. Hibernation: cellular aspects. *Annu. Rev. Physiol.* 41:275–286.

Willis, J. S., and M. Baudysova. 1977. Retention of K^+ in relation to cold resistance of cultured cells from hamsters and human embryos. *Cryobiology* 14:511–515.

Willis, J. S., J. C. Ellory, and M. W. Walowyk. 1980. Temperature sensitivity of the sodium pump in red cells from various hibernator and nonhibernator species. *J. Comp. Physiol.* 138:43–47.

Winter, C. 1973. The influence of temperature on membrane processes. In *Effects of Temperature on Ectothermic Organisms,* ed. W. Weiser. New York: Springer-Verlag, pp. 45–53.

Wood, S. C., G. Lykkeboe, K. Johansen, R. E. Weber, and G. M. O. Maloiy. 1978. Temperature acclimation in the pancake tortoise, *Malacochersus tornieri:* metabolic rate, blood pH, oxygen affinity and red cell organic phosphates. *Comp. Biochem. Physiol.* 59A:155–160.

5. Endothermic Hibernators

Ahlquist, D. A. 1976. Glycerol and alanine metabolism in the hibernating black bear. *The Physiologist* 19:107.

Allweis, C., T. Landau, M. Abeles, and J. Magnes. 1966. The oxidation of uniformly labelled albumin-bound palmitic acid to CO_2 by the perfused cat brain. *J. Neurochem.* 13:795–804.

Atkinson, D. E., and M. N. Camien. 1982. The role of urea synthesis in the removal of metabolic bicarbonate and the regulation of blood pH. *Curr. Top. Cell. Regul.* 21:261–302.

Bailey, E. D., and D. E. Davis. 1965. The utilization of body fat during hibernation in woodchucks. *Can. J. Zool.* 43:701–707.

Bartholomew, G. A. 1981. A matter of size: an examination of endothermy in insects and terrestrial vertebrates. In *Insect Thermoregulation,* ed. B. Heinrich. New York: John Wiley & Sons, pp. 45–78.

Bickler, P. E. 1984. Blood acid-base status of an awake heterothermic rodent, *Spermophilus tereticaudus. Respir. Physiol.* 57:307–316.

Cannon, B., and J. Nedergaard. 1982. The function and properties of brown adipose tissue in the newborn. In *Biochemical Development of the Fetus and Neonate,* ed. C. T. Jones. Amsterdam: Elsevier Biomedical, pp. 697–730.

Chappell, M. A. 1984. Maximum oxygen consumption during exercise and cold exposure in deer mice, *Peromyscus maniculatus Respir. Physiol.* 55:367–377.

Conn, A. R., and R. D. Steele. 1982. Transport of α-keto analogues of amino acids across blood-brain barrier in rats. *Am. J. Physiol.* 243:E272–E277.

Davis, D. E. 1976. Hibernation and circannual rhythms of food consumption in marmots and ground squirrels. *Q. Rev. Biol.* 51:477–514.

Deavers, D. R., and X. J. Musacchia. 1980. Water metabolism and renal function during hibernation and hypothermia. *Fed. Proc.* 39:2969–2973.

Dempster, G., E. I. Grodums, and W. A. Spencer. 1966. Experimental Coxsackie B-3 virus infection in *Citellus lateralis. J. Cell. Physiol.* 67:443–454.

Folk, E., M. Folk, and J. J. Minor. 1972. Physiological condition of three species of bears in winter dens. I.U.C.N. New Series No. 23, Morges, Switzerland: I.U.C.N. Press, pp. 107–124.

Galster, W. A., and P. R. Morrison. 1966. Seasonal changes in serum lipids and proteins in the 13-lined ground squirrel. *Comp. Biochem. Physiol.* 18:489–501.

——— 1975. Gluconeogenesis in arctic ground squirrels between periods of hibernation. *Am. J. Physiol.* 228:325–330.

——— 1976. Seasonal changes in body composition of the arctic ground squirrel, *Citellus undulatus. Can. J. Zool.* 54:74–78.

George, C. P., T. S. Kilduff, F. R. Sharp, and H. C. Heller. 1982. Autoradiographic patterns of hippocampal metabolism during induced hypothermia. *Neurosci. Lett.* 34:233–239.

Grodums, E. I., W. A. Spencer, and G. Dempster. 1966. The hibernation cycle and related changes in the brain fat tissue of *Citellus lateralis. J. Cell. Physiol.* 67:421–430.

Guppy, M., and J. Ballantyne. 1981. The importance of water and oxygen in the evolution of hydrogen shuttle mechanisms. *Comp. Biochem. Physiol.* 69B:1–4.

Hammel, H. T., H. C. Heller, and F. R. Sharp. 1973. Probing the rostral brain stem of anaesthetized, unaesthetized and exercising dogs and of hibernating and euthermic ground squirrels. *Fed. Proc.* 32:1588–1597.

Hand, S. C., and G. N. Somero, 1983. Phosphofructokinase of the hibernator *Citellus beecheyi:* temperature and pH regulation of activity via influences on the tetramer-dimer equilibrium. *Physiol. Zool.* 53:380–389.

Hawkins, R. A., D. W. Williamson, and H. A. Krebs. 1971. Ketone body utilization by adult and suckling rat brain *in vivo. Biochem. J.* 122:13–18.

Hazel, J. R. 1973. The regulation of function by temperature-induced alterations in membrane composition. In *Effects of Temperature on Ectothermic Organisms,* ed. W. Wieser. New York: Springer-Verlag, pp. 55–67.

Heller, H. C., G. W. Colliver, and J. Beard. 1977. Thermoregulation during entrance into hibernation. *Pflügers Arch.* 369:55–59.

Heller, H. C., G. L. Florant, S. F. Glotzbach, and J. M. Walker. 1978. Sleep and torpor—homologous adaptation for energy conservation. In *Dormancy and Developmental Arrest. Experimental Analysis in Plants and Animals,* ed. M. E. Clutter. New York: Academic Press, pp. 270–296.

Hochachka, P. W. 1973. Temperature and pressure adaptation of the binding site of acetylcholine esterase. *Biochem. J.* 143:535–539.

Hochachka, P. W., and G. N. Somero. 1984. *Biochemical Adaptation.* Princeton, N.J.: Princeton University Press.

Hock, R. J. 1960. VIII. Seasonal variation in physiologic functions of arctic ground squirrels and black bears. *Bull. Mus. Comp. Zool.* 124:155–171.

Jameson, E. W. Jr. 1965. Food consumption of hibernating and nonhibernating *Citellus lateralis. J. Mammal.* 46:634–640.

Joel, C. D. 1965. The physiological role of brown adipose tissue. In *Handbook of Physiology,* Section 5, Adipose Tissue, ed. A. E. Renold and G. F. Cahill. Washington, D.C.: American Physiological Society, pp. 59–85.

Kilduff, T. S., F. R. Sharp, and H. C. Heller. 1982. [^{14}C]-2-deoxyglucose uptake in ground squirrel brain during hibernation. *J. Neurosci.* 2:143–157.

——— 1983. Relative 2-deoxyglucose uptake of the paratrigeminal nucleus increases during hibernation. *Brain Res.* 262:117–123.

Klain, G. J., and B. K. Whitten. 1968. Plasma free amino acids in hibernation and arousal. *Comp. Biochem. Physiol.* 27:617–619.

Lehninger, A. L. 1975. *Biochemistry.* New York: Worth Publishers.

Lentz, C. P., and J. S. Hart. 1960. The effect of wind and moisture on heat loss through the fur of newborn caribou. *Can. J. Zool.* 38:679–688.

Lundberg, D. A., R. A. Nelson, H. W. Wahner, and J. D. Jones. 1976. Protein metabolism in the black bear before and during hibernation. *Mayo Clinic Proc.* 51:716–722.

Lyman, C. P. 1958. Oxygen consumption, body temperature and heart rate of woodchucks entering hibernation. *Am. J. Physiol.* 194:83–91.

——— 1965. Circulation in mammalian hibernation. In *Handbook of Physiology,* Section 2, Circulation 3, ed. W. F. Hamilton and P. Dow. Washington, D.C.: American Physiological Society.

Lyman, C. P., and R. C., O'Brien. 1969. Hyperresponsiveness in hibernation. *Symp. S.E.B.* 23:489–509.

Lyman, C. P., J. Willis, A. Malan, and L. Wang. 1982. *Hibernation and Torpor in Mammals and Birds.* New York: Academic Press.

McGarry, S. D., and D. W. Foster. 1980. Regulation of hepatic fatty acid oxidation and ketone body production. *Annu. Rev. Biochem.* 49:395–420.

Malan, A., J. L. Rodeau, and F. Daull. 1981. Intracellular pH in hibernating hamsters. *Cryobiology* 18:100.

Mink, J. W., T. J. Blumenschire, and D. B. Adams. 1981. Ratio of central nervous system to body metabolism in vertebrates: its constancy and functional basis. *Am. J. Physiol.* 241:R203–R212.

Mitchener, G. R. 1977. Effect of climatic condition on the annual activity and hibernation cycle of Richardson's ground squirrels and Columbian ground squirrels. *Can. J. Zool.* 55:693–703.

Morrison, P. R., and W. A. Galster. 1975. Patterns of hibernation in the arctic ground squirrel. *Can. J. Zool.* 53:1345–1355.

Nelson, R. A. 1973. Winter sleep in the black bear: a physiologic and metabolic marvel. *Mayo Clinic Proc.* 48:733–737.

——— 1978. Urea metabolism in the hibernating black bear. *Kidney Int.* 13(Suppl. 8):S177–S179.

——— 1980. Protein and fat metabolism in hibernating bears. *Fed. Proc.* 39:2955–2958.

Nelson, R. A., T. D. I. Beck, and D. L. Steiger. 1984. Ratio of serum urea to serum creatinine in wild black bears. *Science* 226:841–842.

Nelson, R. A., J. D. Jones, H. W. Wahner, D. B. McGill, and C. F. Code. 1975. Nitrogen metabolism in bears: urea metabolism in summer starvation and in winter sleep and role of urinary bladder in water and nitrogen conservation. *Mayo Clinic Proc.* 50:141–146.

Nelson, R. A., H. W. Wahner, J. D. Jones, R. D. Ellefson, and P. E. Zollman. 1973. Metabolism of bears before, during and after winter sleep. *Am. J. Physiol.* 224:491–496.

Neumann, R. L. 1967. Metabolism in the eastern chipmunk (*Tamias striatus*) and the southern flying squirrel (*Glaucomys volans*) during the winter and summer. In *Mammalian Hibernation*, vol. 3, ed. K. C. Fisher, A. R. Dawe, C. P. Lyman, E. Schonbaum, and F. E. South. London: Oliver and Boyd, pp. 64–74.

Newsholme, E. A., and C. Start. 1973. *Regulation in Metabolism.* London: John Wiley & Sons.

Passmore, J. C., E. W. Pfeiffer, and J. R. Templeton. 1975. Urea excretion in the hibernating Columbian ground squirrel *Spermophilus columbianus. J. Exp. Zool.* 192:83–86.

Pengelley, E. T., and K. C. Fisher. 1961. Rhythmical arousal from hibernation in the golden-mantled ground squirrel *Citellus lateralis tesconum. Can J. Zool.* 39:105–120.

Reidesel, M. L., and J. M. Steffen. 1980. Protein metabolism and urea recycling in rodent hibernation. *Fed. Proc.* 39:2959–2963.

Robinson, A., and D. H. Williamson. 1980. Physiological role of ketone bodies as substrates and signals in mammalian tissues. *Physiol. Rev.* 60:145–187.

Smith, R. E., and R. J. Hock. 1963. Brown fat: thermogenic effector of arousal in hibernation. *Science* 140:199–200.

Snap, B. D., and H. C. Heller. 1981. Suppression of metabolism during hibernation in ground squirrels *Citellus lateralis. Physiol Zool.* 54:297–307.

Sokoloff, L. 1977. Relation between physiological function and energy metabolism in the central nervous system. *J. Neurochem.* 29:13–26.

Spencer, W. A., E. I. Grodums, and G. Dempster. 1966. The glyceride fatty acid composition and lipid content of brown and white adipose tissue of the hibernator *Citellus lateralis. J. Cell. Physiol.* 67:431–442.

Strumwasser, F., J. J. Gilliam, and J. L. Smith. 1964. Long-term studies on individual hibernating animals. *Ann. Acad. Sci. Fenn.* Ser. A 4571:399–414.

Swan, H. 1981. Neuroendocrine aspects of hibernation. In *Survival in the Cold: Hibernation and Other Adaptations*, ed. X. J. Musacchia and L. Jansky. Amsterdam: Elsevier North-Holland, pp. 121–138.

Tashima, L. S., S. J. Adelstein, and C. P. Lyman. 1970. Radio glucose utilization by active, hibernating and arousing ground squirrels. *Am. J. Physiol.* 218:303–309.

Tucker, V. A. 1965. The relation between the torpor cycle and heat exchange in the Californian pocket mouse *Perognathus californicus*. *J. Cell. Physiol.* 65:405–414.

Twente, J. W., and J. Twente. 1978. Autonomic regulation of hibernation by *Citellus* and *Eptesicus*. In *Strategies in Cold: Natural Torpidity and Thermogenesis*, ed. L. C. H. Wang and J. W. Hudson. New York: Academic Press, pp. 327–376.

Wang, L. C. H. 1978. Energetic and field aspects of mammalian torpor: the Richardson's ground squirrel. In *Strategies in Cold: Natural Torpidity and Thermogenesis*, ed. L. C. H. Wang and J. W. Hudson. New York: Academic Press, pp. 109–145.

Wang, L. C. H., and J. W. Hudson. 1971. Temperature regulation in normothermic and hibernating eastern chipmunk *Tamius striatus*. *Comp. Biochem. Physiol.* 38A:59–90.

Watts, P. D., N. A. Oritsland, C. Jonkel, and K. Ronald. 1981. Mammalian hibernation and the oxygen consumption of a black bear (*Ursus americanus*). *Comp. Biochem. Physiol.* 69A:121–123.

Whitten, B. K., R. F. Burlington, and M. A. Posiviata. 1974. Temporal changes in amino acid catabolism during arousal from hibernation in the golden-mantled ground squirrel. *Comp. Biochem. Physiol.* 47A:541–546.

Whitten, B. K., and G. S. Klain. 1968. Protein metabolism in hepatic tissue of hibernating and arousing ground squirrels. *Am. J. Physiol.* 214:1360–1362.

Willis, J. S. 1979. Hibernation: cellular aspects. *Annu. Rev. Physiol.* 41:275–286.

Willis, J. S., J. C. Ellery, and M. W. Walowyk. 1980. Temperature sensitivity of the sodium pump in red cells from various hibernator and nonhibernator species. *J. Comp. Physiol.* 138:43–47.

Willis, J. S., S. S. Goldman, and R. J. Foster. 1971. Tissue K concentration in relation to the role of the kidney in hibernation and the cause of periodic arousal. *Comp. Biochem. Physiol.* 39A:437–445.

Wilson, T. L. 1977. Interrelation between pH and temperature for the catalytic rate of the M4 isozyme of lactate dehydrogenase (EC 1.11.27) from goldfish *Carassius auratus*. *Arch. Biochem. Biophys.* 179:378–390.

Zimmermann, M. L. 1982. Carbohydrate and torpor duration in hibernating golden-mantled ground squirrels (*Citellus lateralis*). *J. Comp. Physiol.* 147:129–135.

6. *Estivators*

Boutilier, R. G., D. J. Randall, G. Shelton, and D. P. Toews. 1979. Acid–base relationships in the blood of the toad *Bufo marinus*. *J. Exp. Biol.* 82:357–365.

Dunn, J. F., P. W. Hochachka, W. Davison, and M. Guppy. 1983. Metabolic adjustments to diving and recovery in the African lungfish. *Am. J. Physiol.* 245:R651–R657.

Heatwole, H. 1983. Physiological responses of animals to moisture and temperature. In *Tropical Rain Forest Ecosystems. A. Structure and Function,* ed. F. B. Golley. Amsterdam: Elsevier, pp. 239–265.

Hochachka, P. W. 1980. *Living without Oxygen.* Cambridge, Mass.: Harvard University Press.

Hochachka, P. W., and D. J. Randall. 1978. Water–air breathing transition in vertebrates of the Amazon. *Can. J. Zool.* 56:1–1016.

Hochachka, P. W., and G. N. Somero. 1984. *Biochemical Adaptation.* Princeton, N. J.: Princeton University Press.

Jungreis, A. M. 1978. Insect dormancy. In *Dormancy and Developmental Arrest,* ed. M. E. Clutter. New York: Academic Press, pp. 47–166.

Loveridge, J. P. 1970. Observations on nitrogenous excretion and water relations of *Chiromantis xerampelina. Arnoldia* 5:1–6.

McClanahan, L. Jr., J. N. Stinner, and V. H. Shoemaker. 1978. Skin lipids, water loss, and energy metabolism in a South American tree frog (*Phyllomedusa sauvagei*). *Physiol. Zool.* 51:179–187.

Malan, A. 1978. Intracellular acid–base state at a variable temperature in air-breathing vertebrates and its representation. *Respir. Physiol.* 33:115–119.

Mansingh, A. 1971. Physiological classification of dormancies in insects. *Can. J. Entomol.* 103:983–1009.

Margules, D. L. 1979. β-Endorphin and endoloxone: hormones of the autonomic nervous system for the conservation or expenditure of bodily resources and energy in anticipation of famine or feast. *Neurosci. Biobehav. Rev.* 3:155–162.

Mommsen, T. P., C. J. French, and P. W. Hochachka. 1980. Sites and patterns of protein and amino acid utilization during the spawning migration of salmon. *Can. J. Zool.* 58:1785–1799.

Riddiford, L. M., and J. W. Truman. 1978. Biochemistry of insect hormones and insect growth regulators. In *Biochemistry of Insects,* ed. M. Rockstein. New York: Academic Press, pp. 308–357.

Schmidt-Nielsen, K., C. R. Taylor, and A. Shkolnik. 1971. Desert snails: problems of heat, water, and food. *J. Exp. Biol.* 55:385–398.

Seymour, R. S. 1973. Energy metabolism of dormant spadefoot toads (*Scaphiopus*). *Copeia* 1973, 435–444.

Shoemaker, V. H., and L. R. McClanahan. 1975. Evaporative water loss, nitrogen excretion, and osmoregulation in Phyllomedusine frogs. *J. Comp. Physiol.* 100:331–345.

Simkiss, K. 1968. Calcium and carbonate metabolism in the frog (*Rana temporaria*) during respiratory acidosis. *Am. J. Physiol.* 214:627–634.

Smith, H. W. 1930. Metabolism of the lungfish, *Protopterus aethiopicus. J. Biol. Chem.* 88:97–130.

7. Frozen Insects

Baust, J. G. 1981. Biochemical correlates of cold-hardening in insects. *Cryobiology* 18:186–198.

Baust, J. G., and R. E. Lee. 1982. Environmental triggers to cryoprotectant modulation in separate populations of the gall fly, *Eurosta solidaginis*. *J. Insect Physiol.* 28:431–436.

Beall, P. T. 1983. States of water in biological systems. *Cryobiology* 20:324–334.

Crowe, J. H., L. M. Crowe, and R. Mouradian. 1983. Stabilization of biological membranes at low water activities. *Cryobiology* 20:346–356.

Crowe, J. H., L. M. Crowe, and S. J. O'Dell. 1981. Ice formation during freezing of *Artemia* cysts of variable water contents. *Mol. Physiol.* 1:145–152.

Duman, J. G., and K. Horwath. 1983. The role of hemolymph proteins in the cold tolerance of insects. *Annu. Rev. Physiol.* 45:261–270.

Kuntz, I. B. Jr., and W. Kauzmann. 1974. Hydration of proteins and peptides. *Adv. Protein Chem.* 28:239–345.

Meyer, R. A., M. J. Kushmerick, and T. R. Brown. 1982. Application of ^{31}P-NMR spectroscopy to the study of striated muscle metabolism. *Am. J. Physiol.* 242:C1–C11.

Salt, R. W. 1961. Principles of insect cold-hardiness. *Annu. Rev. Entomol.* 6:55–74.

Storey, K. B. 1983. Metabolism and bound water in overwintering insects. *Cryobiology* 20:365–379.

Storey, K. B., and J. M. Storey. 1983. Biochemistry of freeze tolerance in terrestrial insects. *Trends Biochem. Sci.* 8:242–245.

Storey, K. B., J. G. Baust, and P. Buescher. 1981. Determination of water "bound" by soluble subcellular compartments during low-temperature acclimation in the gall fly larva, *Eurosta solidaginis*. *Cryobiology* 18:315–321.

Storey, K. B., M. Micelli, K. W. Butler, I. C. P. Smith, and R. Deslauriers. 1984. ^{31}P-NMR studies of the freeze tolerant larvae of the gall fly, *Eurosta solidaginis*. *Eur. J. Biochem.* 142:591–595.

8. Frozen Frogs

Farrar, E. S., and R. K. Dupre. 1983. The role of diet in glycogen storage by juvenile bullfrogs prior to overwintering. *Comp. Biochem. Physiol.* 75A:255–260.

Schmid, W. D. 1982. Survival of frogs in low temperature. *Science* 215:697–698.

Storey, K. B. 1985. Freeze tolerance in terrestrial frogs. *Cryo-Letters* 6:115–134.

Storey, K. B., and J. M. Storey. 1984. Biochemical adaptation for freezing tolerance in the wood frog, *Rana sylvatica*. *J. Comp. Physiol.* 155:29–36.

——— 1985. Adaptations of metabolism for freezing tolerance in the gray tree frog, *Hyla versicolor*. *Can. J. Zool.* 63:49–54.

9. Anhydrobiotes

Busa, W. B. 1985. How to succeed at anaerobiosis without really dying. *Mol. Physiol.* 8:351–358.

Busa, W. B., and J. H. Crowe. 1983. Intracellular pH regulates transitions between dormancy and development of brine shrimp (*Artemia salina*) embryos. *Science* 221:366–368.

Busa, W. B., and R. Nuccitelli. 1984. Metabolic regulation via intracellular pH. *Am. J. Physiol* 246:R409–R438.

Clegg, J. S. 1979. Metabolic consequences of the extent and disposition of the aqueous intracellular environment. *J. Exp. Zool.* 215:303–313.

——— 1984. Properties and metabolism of the aqueous cytoplasm and its boundaries. *Am. J. Physiol.* 246:R133–R151.

Crowe, J. H. 1971. Anhydrobiosis: an unsolved problem. *Am. Naturalist* 105:563–573.

Crowe, J. H., and J. S. Clegg, eds. 1978. *Dry Biological Systems.* New York: Academic Press.

Hickernell, L. M. 1917. A study of desiccation in the rotifer, *Philodina roseola,* with special reference to cytological changes accompanying desiccation. *Biol. Bull.* 32:343–397.

Hinton, H. E. 1968. Reversible suspension of metabolism and the origin of life. *Proc. R. Soc. Lond.* Ser. B 171:43–57.

Keilin, D. 1959. The problem of anabiosis or latent life: history and current concepts. *Proc. R. Soc. Lond.* Ser. B 150:149–191.

Persoone, G., P. Sorgeloos, A. Roels, and E. Jaspers, eds. 1980. *The Brine Shrimp Artemia.* Wetteren, Belgium: Universal Press, 3 volumes.

10. Perspectives

Andersson, B. S., and D. P. Jones. 1985. Ion distribution in hepatocytes during anoxia. Int. Union Biochemistry, ABSTRACTS FR-465.

Arrhenius, S. 1908. *Worlds in the Making.* New York: Harper & Row.

Childress, J. J., and G. N. Somero. 1979. Depth-related enzyme activities in muscle, brain, and heart of deep-living pelagic marine teleosts. *Mar. Biol.* 52:273–283.

Crick, F. H. C. 1981. *Life Itself,* Austin, Texas: S & X Press.

Crick, F. H. C., and L. E. Orgel. 1973. Directed panspermia. *Icarus* 19:341–346.

Crowe, J. H., L. M. Crowe, and R. Mauradian. 1983. Stabilization of biological membranes at low water activities. *Cryobiology* 20:346–356.

Epstein, F. H., M. Brezis, P. Silva, and S. Rosen. 1985. Protection against hypoxic injury in renal medulla. *Mol. Physiol.* 8:525–534.

Fox, S. W., and K. Dose. 1977. *Molecular Evolution and the Origin of Life.* New York: Marcel Dekker.

Haldane, J. B. S. 1929. *The Origin of Life.* Reprinted in J. D. Bernal, *The Origin of Life,* Cleveland: World, 1967.

Hanson, R. S. 1976. Dormant and resistant stages of procaryotic cells. In *Chemical Evolution of the Giant Planets,* ed. C. Ponnamperuma. New York: Academic Press, pp. 107–120.

Hoyle, F., and C. Wickramasinghe. 1981. *Space Travellers, the Bringers of Life.* Cardiff: University of Cardiff Press.

Jones, D. P., F. G. Kennedy, B. S. Andersson, T. Y. Aw, and E. Wilson. 1985. When is a mammalian cell hypoxic? Insights from studies of cells *vs* mitochondria. *Mol. Physiol.* 8:473–482.

Margules, D. L. 1979. β-Endorphin and endoloxone: hormones of the autonomic nervous system for the conservation or expenditure of bodily resources and energy in anticipation of famine or feast. *Neurosci. Biobehav. Rev.* 3:155–162.

Oparin, A. I. 1924. *The Origin of Life.* Moscow: Proiskhozhdenie Zhizny, Izd. Moskovski Rabochii. Tr. in J. D. Bernal, *The Origin of Life,* Cleveland: World, 1967.

Robinson, R. 1966. The origins of petroleum. *Nature* 212:1291–1295.

——— 1967. Origins of oil, a correction and further comment on the Brunnock even C-number predominance in certain higher alkanes of African crude and on the biogenesis of nonacosane. *Nature* 214:263.

Saez, L., T. Zuvic, R. Amthauer, E. Rodriguez, and M. Krauskopf. 1984. Fish liver protein synthesis during cold acclimatization: seasonal changes of the ultrastructure of the carp hepatocyte. *J. Exp. Zool.* 230:175–186.

Sussman, A. S., and H. O. Hàlvorson. 1966. *Spores: Their Dormancy and Germination.* New York: Harper & Row.

Whittow, G. C., C. A. Scammell, J. K. Manuel, D. Rand, and M. Leong. 1977. Temperature regulation in a hypometabolic primate, the slow loris. *Arch. Intl. Physiol. Biochim.* 85:139–151.

Appendix A. Interactions between O_2 and Metabolism

Benson, D. M., J. A. Knopp, and I. S. Longmuir. 1980. Intracellular O_2 measurements of mouse liver cells using quantitative fluorescence video microscopy. *Biochim. Biophys. Acta* 591:187–197.

Chance, B. 1976. Pyridine nucleotide as an indicator of the O_2 requirements for energy-linked functions of mitochondria. *Circ. Res.* 38:131–138.

Connett, R. J., T. E. J. Gayeski, and C. R. Honig. 1984. Lactate accumulation in fully aerobic working dog gracilis muscle. *Am. J. Physiol.* 246:H120–H128.

Edelstone, D. I., M. E. Paulone, and I. R. Holzman. 1984. Hepatic oxygenation during arterial hypoxemia in neonatal lambs. *Am. J. Obstet. Gynecol.* 150:513–518.

Grubb, B., and G. E. Folk, Jr. 1978. Skeletal muscle $\dot{V}_{O_2}$ in rat and lemming: effect of blood flow rate. *J. Comp. Physiol.* 128:185–188.

Hammerstedt, R. H., and H. A. Lardy. 1983. The effect of substrate cycling on the ATP yield of sperm glycolysis. *J. Biol. Chem.* 258:8759–8765.

Hochachka, P. W. 1982a. Metabolic arrest as a mechanism of protection against hypoxia. In *Protection of Tissues against Hypoxia,* ed. A. Wauquier, M. Borgers, and W. K. Avery. Amsterdam: Elsevier Biomedical, pp. 1–12.

——— 1982b. Anaerobic metabolism: living without oxygen. In *A Companion to Animal Physiology*, ed. C. R. Taylor, K. Johansen, and L. Bolis. Cambridge: Cambridge University Press, pp. 138–150.

Hochachka, P. W., and J. F. Dunn. 1983. Metabolic arrest: the most effective means of protection against hypoxia. In *Hypoxia, Exercise, and Altitude*, ed. J. R. Sutton, C. S. Houston, and N. L. Jones. New York: Alan R. Liss, pp. 297–310.

Honig, C. R., T. E. J. Gayeski, W. Federspiel, A. Clark, Jr., and P. Clark. 1984. Muscle O_2 gradients from hemoglobin to cytochrome: new concepts, new complexities. *Adv. Exp. Med. Biol.* 169:23–38.

Horowitz, S. B., and T. W. Pearson. 1981. Intracellular monosaccharide and amino acid concentrations and activities and the mechanism of insulin action. *Mol. Cell. Biol.* 1:769–784.

Jones, D. P. 1984. Effect of mitochondrial clustering on O_2 supply in hepatocytes. *Am. J. Physiol.* 247:C83–C89.

Jones, D. P., and F. G. Kennedy. 1982. Intracellular O_2 gradients in cardiac myocytes. Lack of a role for myoglobin in facilitation of intracellular O_2 diffusion. *Biochem. Biophys. Res. Commun.* 105:419–424.

——— 1986. Analysis of intracellular oxygenation of isolated adult cardiac myocytes. *Am. J. Physiol.* 250:C384–C390.

Kennedy, F. G., and D. P. Jones. 1986. Oxygen dependence of mitochondrial function in isolated rat cardiac myocytes. *Am. J. Physiol.* 250:C374–C383.

Kinter, D., J. H. Fitzpatrick, Jr., J. A. Louie, and D. D. Gilboe. 1984. Cerebral O_2 and energy metabolism during and after 30 minutes of moderate hypoxia. *Am. J. Physiol.* 247:E475–E482.

Lutz, J., H. Heinrich, and E. Bauereisen. 1975. O_2 supply and uptake in the liver and the intestine. *Pflügers Arch.* 360:7–15.

Mastro, A. M., M. A. Babich, W. D. Taylor, and A. D. Keith. 1984. Diffusion of a small molecule in the cytoplasm of mammalian cells. *Proc. Natl. Acad. Sci. U.S.A.* 81:3414–3418.

Mulligan, E., and S. Lahiri. 1981. Mitochondrial oxidative metabolism and chemoreception in the carotid body. In *Arterial Chemoreceptors*. Leicester: Leicester University Press, pp. 316–327.

Nuutinen, E. M., K. Nishiki, M. Erecinska, and D. F. Wilson. 1982. Role of mitochondrial oxidative phosphorylation in regulation of coronary blood flow. *Am. J. Physiol.* 243:H159–H169.

Paine, P. L., and S. B. Horowitz. 1980. The movement of material between nucleus and cytoplasm. *Cell Biol.* 4:299–328.

Rosenmann, M., and P. Morrison. 1974. Physiological responses to hypoxia in the tundra vole. *Am. J. Physiol.* 227:734–739.

Siesjo, B. K. 1978. *Brain Energy Metabolism*. New York: John Wiley & Sons.

Whalen, W. J., D. Buerk, and C. A. Thuning. 1973. Blood flow-limited oxygen consumption in resting cat skeletal muscle. *Am. J. Physiol.* 224:763–768.

Wilson, D. F., C. S. Owen, and M. Erecinska. 1979. Quantitative dependence of mitochondrial oxidative phosphorylation on O_2 concentration: a mathematical model. *Arch. Biochem. Biophys.* 195:494–504.

Appendix B. Normal and Reversed Pasteur Effects

Bagnasco, S., D. Good, R. Balaban, and M. Burg. 1985. Lactate production in isolated segments of the rat nephron. *Am. J. Physiol.* 249:F522–F526.

Cannata, J. J. B., and J. J. Cazzulo. 1984. The aerobic fermentation of glucose by *Trypanosoma cruzi. Comp. Biochem. Physiol.* 79B:297–308.

Clarke, F. M., P. Stephan, G. Huxham, D. Hamilton, and D. J. Morton. 1984. Metabolic dependence of glycolytic enzyme binding in rat and sheep heart. *Eur. J. Biochem.* 138:643–649.

Clegg, J. S. 1984. Properties and metabolism of the aqueous cytoplasm and its boundaries. *Am. J. Physiol.* 246:R133–R151.

Kacser, H., and J. A. Burns. 1973. The control of flux. *Symp. Soc. Exp. Biol.* 32:65–104.

Kempner, E. S., and J. H. Miller. 1968. The molecular biology of *Euglena gracilis.* V. Enzyme localization. *Exp. Cell Res.* 51:150–156.

Lynch, R. M., and R. J. Paul. 1983. Compartmentation of glycolytic and glycogenolytic metabolism in vascular smooth muscle. *Science* 222:1344–1346.

Pasteur, L. 1861. Experiences et vues nouvelles sur la nature des fermentations. *C. R. Acad. Sci.* 52:1260–1264.

Paul, R. J. 1983. Functional compartmentalization of oxidative and glycolytic metabolism in vascular smooth muscle. *Am. J. Physiol.* 244:C399–C409.

Racker, E. 1976. *A New Look at Mechanisms in Bioenergetics.* New York: Academic Press.

Rapaport, T. A., R. Heinrich, and S. M. Rapaport. 1976. A minimal comprehensive model describing steady states, quasi-steady states, and time-dependent processes. *Biochem. J.* 154:449–469.

Regen, D. M., and S. J. Pilkis. 1984. Sensitivity of pathway rate to activities of substrate-cycle enzymes: application to gluconeogenesis and glycolysis. *J. Theor. Biol.* 111:635–658.

Reibstein, D., J. A. den Hollander, S. J. Pilkis, and R. G. Shulman. 1986. Studies on the regulation of yeast phosphofructo-1-kinase: its role in aerobic and anaerobic glycolysis. *Biochemistry* 25:219–227.

Srivastava, D. K., and Bernhard, S. A. 1986. Enzyme-enzyme interactions and the regulation of metabolic reaction pathways. *Curr. Top. Cell. Regul.* 28: in press.

Ureta, T. 1978. The role of isozymes in metabolism: a model of metabolic pathways as the basis for the biological role of isozymes. *Curr. Top. Cell. Regul.* 13:233–266.

Yoshino, M., and K. Murakami. 1983. Role of AMP deaminase in the response of phosphofructokinase to the adenylate energy charge. *Biochem. Biophys. Res. Commun.* 112:96–101.

Index